In this series original contributions of mathematical research in all fields are contained, such as

— research monographs
— collections of papers to a single topic
— reports on congresses of exceptional interest for mathematical research.

This series is aimed at promoting quick information and communication between mathematicians of the various special branches.

In diese Reihe werden Originalbeiträge zu allen Gebieten der mathematischen Forschung aufgenommen wie

— Forschungsmonographien
— Sammlungen von Arbeiten zu einem speziellen Thema
— Berichte von Tagungen, die für die mathematische Forschung besonders aktuell sind.

Die Reihe soll die schnelle Information und gute Kommunikation zwischen den Mathematikern der verschiedenen Fachgebiete fördern.

Manuscripts
in English and German comprising at least 100 pages and not more than 500 pages can be admitted to this series. With respect to a quick publication the manuscripts are reproduced photomechanically. Authors who are interested in this series please turn directly to the 'Akademie-Verlag'. Here you will get more detailed information about the form of the manuscripts and the modalities of publication.

Manuskripte
in englischer und deutscher Sprache, die mindestens 100 Seiten und nicht mehr als 500 Seiten umfassen, können in diese Reihe aufgenommen werden. Im Interesse einer schnellen Publikation werden die Manuskripte auf fotomechanischem Weg reproduziert. Autoren, die an der Veröffentlichung entsprechender Arbeiten in dieser Reihe interessiert sind, wenden sich bitte direkt an den Akademie-Verlag. Sie erhalten dort genauere Informationen über die Gestaltung der Manuskripte und die Modalitäten der Veröffentlichung.

A. Meyer

Modern Algorithms for Large Sparse Eigenvalue Problems

Mathematical Research · Mathematische Forschung

Wissenschaftliche Beiträge
herausgegeben von der
Akademie der Wissenschaften der DDR
Karl-Weierstraß-Institut für Mathematik

Band 34

Modern Algorithms for Large Sparse Eigenvalue Problems

by
A. Meyer

Modern Algorithms for Large Sparse Eigenvalue Problems

by

Arnd Meyer

Akademie-Verlag Berlin 1987

Autor:
Dr. Arnd Meyer
Technische Hochschule Karl-Marx-Stadt
Sektion Mathematik
PF 964
Karl-Marx-Stadt
DDR 9010

Die Titel dieser Reihe werden vom Originalmanuskript der Autoren reproduziert

ISBN 3-05-500332-2

ISSN 0138-3019
Erschienen im Akademie-Verlag Berlin
DDR-1086 Berlin, Leipziger Straße 3-4

Lizenznummer: 202 · 100/415/87
Printed in the German Democratic Republic
Gesamtherstellung: VEB Kongeß- und Werbedruck, 9273 Oberlungwitz
Lektor: Dr. Reinhard Höppner
LSV 1065
Bestellnummer: 763 712 1 (2182/34)
01800

PREFACE

The problem of the extraction of a few eigenvalues with corresponding eigenvectors from large sparse matrices occurs in many branches of science and engineering. It is common knowledge that especially the investigation of the dynamical behaviour of structures requires such computations for a generalized eigenvalue problem, whose matrices are the results of discretization methods.

This volume presents some algorithms for the calculation of a few eigenvalues and eigenvectors of large sparse problems. On the one hand, some well-known methods such as Simultaneous Iteration and the Lanczos-algorithm are described, which are compared with a class of simultaneous methods with gradient-type directions of movement.

For the latter class (less frequently treated in the literature), the properties of convergence are clarified and the application to vibration analysis of structures, using modern preconditionings, are discussed.

This book is written for advanced undergraduates and graduates, who specialize in numerical linear algebra, as well as for mathematically inclined engineers and scientists.

The author has occasion to thank his colleagues Prof. V. Friedrich, Dr. W. Lang, Dr. B. Döhler as well as Prof. I. Marek for the many pieces of helpful advice and for fruitful discussions on this topic. Furthermore, the author wishes to thank Mrs. I. Scholz for her excellent typing work and Mr. N. Peshier.

The author welcomes in advance any comments and critical remarks as to the present publication.

Karl-Marx-Stadt
March 1986 Arnd Meyer

CONTENTS

NOTATIONS

$\alpha, \beta, \gamma, \ldots$	real (complex) numbers
$\mathbf{R}(\mathbf{C})$	the field of real (complex) numbers
$[\alpha, \beta]$	closed interval ($\alpha, \beta \in \mathbf{R}$)
i	imaginary unit
$a, b, \ldots, t$	vectors in $\mathbf{R}^p$, $t = (\tau_j)$
A, B, E, F, Q, T, U	matrices in $\mathbf{R}^{p \times q}$, $A = (\alpha_{ij})$
Λ, Δ	diagonal matrices
x, y, z, r, s	elements of $\mathbf{H}$, a finite dimensional space with inner product $\langle ., . \rangle$
X, Y, Z, R, S, W	elements of $\mathbf{H}^p$
$\mathcal{A}, \mathcal{B}, \mathcal{C}, \mathcal{P}, \ldots$	linear operators mapping $\mathbf{H}$ into itself
λ_i	eigenvalues) of $\mathcal{A}$ or of $\{\mathcal{A}, \mathcal{B}\}$
φ_i	eigenvectors) of $\mathcal{A}$ or of $\{\mathcal{A}, \mathcal{B}\}$
K	(nxn) stiffness matrix
M	(nxn) mass matrix
C	(nxn) preconditioning
ω_i^2	eigenvalues of $\{K, M\}$

1. INTRODUCTION

This volume presents some methods for the calculation of a few eigenvalues and eigenvectors of large sparse matrix eigenvalue problems. Such problems arise in many branches of science and engineering. Above all they appear at the heart of analyses of structures. Here, an (nxn) global stiffness matrix K and an (nxn) global mass matrix M are generated using finite element or finite difference methods. Both matrices are symmetric and real, at least K is positive definite. In most cases a few of the smallest positive eigenvalues ω_i^2 of the problem

$$K \varphi_i = \omega_i^2 M \varphi_i \tag{1.1}$$

are to be calculated together with their eigenvectors φ_i. The eigenvalues ω_i correspond to the natural frequencies and the vectors φ_i determine the shapes of the fundamental modes of the vibration. Some other applications are known from quantum chemistry (see [SHAVITT 1, DAVIDSON 1]), plasma and reactor physics (see [PARLETT 2, FRACH/ GREINER-MAI/ SUSCHOWK 1]).

The methods for solving eigenvalue problems can be divided into two main classes. For "small" problems, where at least one full dense (nxn)-matrix can be held in the fast store of a computer, the algorithms of [EISPACK 1 and 2] are sufficient for all practical calculations. We will deal with the other class of problems, where the dimension n is large enough, suggesting that one should cherish the large number of zero elements of the matrices. This field of sparse matrix calculations has been widely investigated since the seventies and is still under development.

For the computation of a few eigenvalues and corresponding eigenvectors of the problem (1.1) two algorithms have been mainly considered in the literature and have found an extensive application in current F.E.M. packages for dynamical research. These are on the one hand the Simultaneous Iteration (or Subspace Iteration) (see Chapter 3.5) and the Lanczos-algorithms on the other hand (Chapter 3.6). As a general problem both algorithms require the transformation of the original generalized eigenvalue problem (1.1) into a special one

$$\mathcal{A} \varphi_i = \lambda_i \varphi_i \tag{1.2}$$

obviously with the help of the substitution

$$\mathcal{A} = K^{-1}M, \quad \lambda_i = \omega_i^{-2}.$$

The matrix operator $\mathcal{A}$ will never be calculated (it would be a dense (nxn)-matrix), moreover, each step of these algorithms performs a calculation of $\mathcal{A}x$ for some given vectors x, hence, a multiplication by M and (the difficulty) the solution of linear systems with the stiffness matrix. If the original problem (1.1) easily permits such solutions (for instance, whenever a factorization of K is possible), then Lanczos-algorithms are superior to the Simultaneous Iteration in most cases.

If the user cannot or will not compute factorizations of K, the application of these methods becomes far more complicated. This will especially occur for future developments with very fine discretizations for improving the mathematical models.

One way-out would consist in an iterative procedure for obtaining the solution of the linear systems in using the above algorithms. Here, unfortunately, the Lanczos-algorithms would require a very accurate solution at each step. Each solution is orthogonal (with respect to a certain inner product) to all previous ones, hence, we never dispose of a suitable starting vector for decreasing this inner iteration number. On the other hand, the Simultaneous Iteration could decrease this number of iterations due to the facts that for earlier steps a rough solution will be sufficient (i.e. we obtain $y \approx \mathcal{A}x$) and for later steps the solutions are the convergent eigenvector approximations, hence, good starting vectors are available. This advantage is kept by our new algorithms, which also act with $\mathcal{A} = K^{-1}M$ and accelerate the Simultaneous Iteration substantially.

Another way-out consists in the construction of alternative algorithms for the eigenvalue problem, which do not require solutions of linear systems with K.

The author's main concern is to present a general class of algorithms for the generalized eigenvalue problem, which are based on a gradient-type direction of movement and on a Rayleigh-Ritz-procedure for choosing appropriate subspaces. The general approach investigates algorithms for the eigenvalue problem

$$\mathcal{A}\varphi = \lambda \mathcal{B}\varphi$$

of symmetric operators $\mathcal{A}$ and $\mathcal{B}$ (with respect to a proper inner product $\langle x,y\rangle = y^T\mathcal{F}x$). Here, we are able to introduce a free preconditioning matrix C by the substitutions

$$\mathcal{A} = C^{-1}M, \quad \mathcal{B} = C^{-1}K, \quad \lambda_1 = \omega_1^{-2}, \quad \mathcal{F} = C ,$$

which leads to an algorithm that does not require the solution of large sparse linear systems with K or M (with C only). Hence, we take advantage of the sparsity of K and M.

New results of the convergence of these algorithms are published here for the first time and may be resorted to when choosing C. The convergence properties will continuously depend on C (up to the extreme choice $C = K$, where the operator $\mathcal{A} = K^{-1}M$ as for Simultaneous Iteration or for the Lanczos-algorithm arises).

This special case $\mathcal{A} = K^{-1}M$ will be widely discussed in the present volume. We obtain a symmetric operator $\mathcal{A}$ in defining the inner product matrix $\mathcal{F}$. The freedom of an arbitrary function f in this definition

$$\mathcal{F} = K\ f(K^{-1}M)$$

is used for the construction of exceedingly cheap implementations of the basic algorithms. Furthermore, the concept of spectral transformation is also applicable to the resulting method by changing $\mathcal{A}$ into another function of $\mathcal{A}$.

The algorithms described in this volume have been extensively applied to various practical problems in machine dynamics and reactor physics, but due to the limited size this publication is mainly devoted to the theoretical foundations of the methods. More profound assertions to the comparison of the methods can be found in [MEYER 3].

In order to calculate a few eigenvalues and corresponding eigenvectors of large sparse problems the idea of a simultaneous calculation has been found to be much superior due to at least two reasons. At first an acceleration of convergence arises, secondly, each step of the iteration processes needs, for instance, the computation of p matrix-vector-products $\mathcal{A}x_1, \ldots, \mathcal{A}x_p$ rather than of one product only. If the large matrices have been stored on secondary storage, both calculations require the same read-in of all information about $\mathcal{A}$.

Hence, the simultaneous use of p vectors decreases the data-transfer by a factor of at least p. Thus, we will mainly deal with simultaneous algorithms. For their development and investigation we introduce some basic tools in Chapter 2, as the product space $\mathbf{H}^p$ of all p-tuples of elements from $\mathbf{H}$ and the functional $\mu(X)$, $X \in \mathbf{H}^p$.

2. BASIC RELATIONS AND NOTATION

This chapter contains a brief description of the basic relations and basic tools used in the construction and investigation of simultaneous algorithms for computing some eigenvalues and corresponding eigenvectors of a generalized eigenvalue problem. Some well-known results on Ritz-projections into smaller subspaces are also presented.

2.1. The Product Space $\mathbb{H}^p$

Let $\mathbb{H}$ be an inner product space of finite dimension n with the elements x, y, z (furthermore r, s, w). The inner product in $\mathbb{H}$ is denoted by

$$\langle x,y\rangle = y^H x, \qquad x,y \in \mathbb{H}.$$

We restrict here to the real case, but a generalization to a complex space $\mathbb{H}$ is straightforward.

To apply our algorithms to large dimensional eigenvalue problems, we set $\mathbb{H}$ to be the n-dimensional real vector space $\mathbb{R}^n$, but generally not having the usual Euclidian inner product. That is, in case of application we define

$$y^H x = y^T \mathbb{F} x$$

for a special symmetric positive definite (nxn)- matrix $\mathbb{F}$.

By $\mathbb{H}^p$ we define the product space of all p-tuples

$$X = (x_1,\dots,x_p)$$

of elements $x_1,\dots,x_p \in \mathbb{H}$. The elements of $\mathbb{H}^p$ will be denoted by X, Y, Z (furthermore R, S, W). In our application X is a rectangular matrix with the columns $x_1,\dots,x_p$, so we generally call x_i the i-th column of X. All other capital letters (especially A, B, C, ..., Q, T and U) denote (pxq)-matrices and $\mathbb{R}^{pxq}$ the space of all (pxq)-matrices. Analogously a, b, q, u, t denote vectors in $\mathbb{R}^p$

$$a = \begin{pmatrix} \alpha_1 \\ \vdots \\ \alpha_p \end{pmatrix}, \qquad \alpha_i \in \mathbb{R},$$

$$A = (\alpha_{ij}), \qquad \alpha_{ij} \in \mathbb{R}.$$

With the definition

$$\langle X,Y\rangle = \sum_{i=1}^{p} \langle x_i,y_i\rangle$$

the space $\mathbb{H}^p$ is also an inner product space.

According to the usual matrix multiplication, we define

$$y = Xt = \sum_{i=1}^{p} \tau_i x_i \in \mathbb{H}$$

for all vectors $t = (\tau_i) \in \mathbb{R}^p$. In the same way,

$$Y = XT$$

represents an element $Y = (y_1,\dots y_q) \in \mathbb{H}^q$ for any (pxq)-matrix $T = (\tau_{ij})$, $i = 1,\dots,p$; $j = 1,\dots,q$, where

$$y_j = \sum_{i=1}^{p} \tau_{ij} x_i, \quad j = 1,\dots,q.$$

We denote the linear operators from $\mathbb{H}$ into $\mathbb{H}$ with $\mathcal{A}, \mathcal{B}, \mathcal{C}, \dots$

With the definition

$$\mathcal{A} X = (\mathcal{A} x_1, \dots, \mathcal{A} x_p) \in \mathbb{H}^p$$

they also map $\mathbb{H}^p$ into itself.
For the specification of simultaneous algorithms matrices are used consisting in inner products of all columns of two elements $X \in \mathbb{H}^p$ and $Y \in \mathbb{H}^q$. Here, we define

$$A = Y^H X \in \mathbb{R}^{q \times p}$$

to be the matrix with elements $\alpha_{ij} = y_i^H x_j = \langle x_j, y_i \rangle$ for all $i = 1,\dots,q$ and $j = 1,\dots,p$.
Usually, this is a bilinear mapping from $\mathbb{H}^p \times \mathbb{H}^q$ into $\mathbb{R}^{q \times p}$. Furthermore note, that the previous definitions lead to the following corollaries:

1. $Y^H X = (X^H Y)^T$,

2. $Y^H(\mathcal{A} X) = (\mathcal{A}^* Y)^H X$,

3. $Y^H(XT) = (Y^H X)T$,

4. $(YU)^H X = U^T Y^H X$

for all $X \in \mathbb{H}^p$, $Y \in \mathbb{H}^q$, $T \in \mathbb{R}^{p \times m}$, $U \in \mathbb{R}^{q \times m}$.
Here, A^T denotes the transpose of a matrix A and $\mathcal{A}^*$ is the adjoint operator of $\mathcal{A}$:

$$\langle \mathcal{A} x, y \rangle = \langle x, \mathcal{A}^* y \rangle, \quad \forall x, y \in \mathbb{H}.$$

Our definition of inner products in $\mathbb{H}$ and $\mathbb{H}^p$ produces norms in these spaces, which will be written without any index

$$\|x\| = \langle x, x \rangle^{\frac{1}{2}}, \qquad x \in \mathbb{H},$$

$$\|X\| = \langle X,X\rangle^{\frac{1}{2}} = (\operatorname{tr} X^H X)^{\frac{1}{2}}, \quad X \in \mathbb{H}^p.$$

Here, tr denotes the trace of the matrix $X^H X$. The linear operator $\mathcal{A}$ mapping $\mathbb{H}$ into itself has the norm

$$\|\mathcal{A}\| = \max_{\|X\|=1} \|\mathcal{A} X\|$$

Furthermore some norms of vectors in $\mathbb{R}^p$ and of (pxq)-matrices will be needed. These are distinguished from above by indices:

$$\|a\|_2 = (a^T a)^{\frac{1}{2}} = (\sum_{i=1}^{p} \alpha_i^2)^{\frac{1}{2}} ,$$

the Euclidian vector norm for $a \in \mathbb{R}^p$,

$$\|A\|_2 = (\varrho(A^T A))^{\frac{1}{2}} = \max_{\|a\|_2=1} \|Aa\|_2 ,$$

the spectral norm of a matrix $A \in \mathbb{R}^{pxq}$ ($\varrho(A)$ denotes the spectral radius of A),

$$\|A\|_F = (\operatorname{tr} A^T A)^{\frac{1}{2}} = (\sum \alpha_{ij}^2)^{\frac{1}{2}} ,$$

the Frobenius norm of the matrix $A = (\alpha_{ij})$.

There are some important relations between norms of the matrixes $X^H Y$ and the elements X and Y. All these relations are listed in the following lemma, for proofs see [MEYER 1].

Note that the matrix $X^H X$ is nonnegative definite symmetric, thus

$$\varrho(X^H X) = \|X^H X\|_2 .$$

<u>Lemma 2.1.1.</u> For all $X \in \mathbb{H}^p$, $Y \in \mathbb{H}^q$, matrices $T \in \mathbb{R}^{pxs}$, vectors $t \in \mathbb{R}^p$ and operators $\mathcal{A}$ mapping $\mathbb{H}$ into itself the following relations are valid:

$$\|XT\| \leq \|X\| \|T\|_2 \tag{2.1.1}$$

$$\|XT\|^2 \leq \|X^H X\|_2 \; \|T\|_F^2 \tag{2.1.2}$$

$$\|Xt\|^2 \leq \|X^H X\|_2 \; \|t\|_2^2 \tag{2.1.3}$$

$$\|X^H Y\|_2^2 \leq \|X^H X\|_2 \; \|Y^H Y\|_2 \tag{2.1.4}$$

$$\|X^H Y\|_F^2 \leq \|X^H X\|_2 \; \|Y\|^2 \tag{2.1.5}$$

$$\|\mathcal{A} X\| \leq \|\mathcal{A}\| \|X\| \tag{2.1.6}$$

If the operator $\mathcal{A}$ is symmetric with

$$\beta_1 \cdot \langle x,x\rangle \leq \langle \mathcal{A} x,x\rangle \leq \beta_2 \langle x,x\rangle ,$$

then

$$\beta_1 \|X\|^2 \leq \operatorname{tr} X^H \mathcal{A} X \leq \beta_2 \|X\|^2, \tag{2.1.7}$$

and if $\beta_2 > 0$, then

$$\beta_1 \|X^H X\|_2 \leq \|X^H \mathcal{A} X\|_2 \leq \beta_2 \|X^H X\|_2 \qquad (2.1.8)$$

For later definitions and proofs the function trA for arbitrary square matrices $A \in \mathbb{R}^{p \times p}$ renders an important tool, thus we add here some well-known trace properties and trace inequalities.

<u>Lemma 2.1.2.</u> For all matrices A and $B \in \mathbb{R}^{p \times p}$ the following relations are valid:

$$\operatorname{tr} AB = \operatorname{tr} BA \qquad (2.1.9)$$

(even for rectangular matrices of proper dimensions)

$$\operatorname{tr} A = \operatorname{tr} A^T \qquad (2.1.10)$$

(thus $\operatorname{tr} AB = \operatorname{tr} AB^T$ for symmetric A)

$$\operatorname{tr} A = \sum_{i=1}^{p} \delta_i,$$

where δ_i denote the p eigenvalues of A.

<u>Lemma 2.1.3.</u> If A, B, C are symmetric (pxp)-matrices with

$$t^T At \leq t^T Bt \quad \text{and}$$

$$t^T Ct \geq 0 \qquad \forall t \in \mathbb{R}^p$$

then

$$\operatorname{tr} CA \leq \operatorname{tr} CB$$

Proof. [MEYER 1]

<u>Corollary.</u>

$$\lambda_{min}(A) \cdot \operatorname{tr} C \leq \operatorname{tr} CA \leq \lambda_{max}(A) \cdot \operatorname{tr} C$$

for symmetric A and C and nonnegative definite C.

2.2. The Generalized Eigenvalue Problem in $\mathbb{H}$

Throughout this paper we deal with symmetric operators $\mathcal{A}$ and $\mathcal{B}$, which map $\mathbb{H}$ into itself. Additional $\mathcal{B}$ be positive definite. We investigate algorithms for the simultaneous calculation of the p largest eigenvalues (with corresponding eigenvectors) of the generalized eigenvalue problem

$$\mathcal{A} x = \lambda \mathcal{B} x. \qquad (2.2.1)$$

Thus,

$$\lambda_1 \geq \lambda_2 \geq \ldots \geq \lambda_n \qquad (2.2.2)$$

denote the n eigenvalues of the problem (2.2.1), if we count them according to multiplicities. We denote the eigenvectors by φ_i to emphasize these special elements in $\mathbb{H}$. Hence,

$$\mathcal{A} \varphi_i = \lambda_i \mathcal{B} \varphi_i, \qquad i = 1, \ldots, n, \qquad (2.2.3)$$

with the well-known orthogonalities

$$\langle \mathcal{A} \varphi_i, \varphi_j \rangle = \langle \mathcal{B} \varphi_i, \varphi_j \rangle = 0 \quad \forall i \neq j \qquad (2.2.4)$$

(for multiple eigenvalues the eigenvectors may be suitably chosen to fulfil (2.2.4) for all i and j).
Furthermore, let all eigenvectors be $\mathcal{B}$-normalized

$$\langle \mathcal{B}\varphi_i, \varphi_i \rangle = 1, \quad i = 1,\dots,n.$$

Our aim is the calculation of the first p eigenvectors, so we collect them to the element

$$\Phi_1 = \Phi_1^{(p)} = (\varphi_1,\dots,\varphi_p) \in \mathbb{H}^p.$$

With the notation

$$\Lambda_1 = \Lambda_1^{(p)} = \operatorname{diag}(\lambda_1,\dots,\lambda_p) \in \mathbb{R}^{p\times p},$$

$$\Lambda_2 = \Lambda_2^{(p)} = \operatorname{diag}(\lambda_{p+1},\dots,\lambda_n) \in \mathbb{R}^{(n-p)\times(n-p)}$$

and

$$\Phi_2 = \Phi_2^{(p)} = (\varphi_{p+1},\dots,\varphi_n) \in \mathbb{H}^{n-p}$$

we write (2.2.3) as

$$\mathcal{A}\,\Phi_1 = \mathcal{B}\,\Phi_1\Lambda_1$$

resp.

$$\mathcal{A}\,\Phi_2 = \mathcal{B}\,\Phi_2\Lambda_2 \tag{2.2.5}$$

furthermore, the orthonormality conditions are

$$\Phi_1^H\mathcal{B}\,\Phi_2 = \Phi_1^H\mathcal{A}\,\Phi_2 = \mathbb{O},$$

$$\Phi_1^H\mathcal{B}\,\Phi_1 = I \qquad \text{((pxp)-identity matrix)},$$

$$\Phi_2^H\mathcal{B}\,\Phi_2 = I \qquad ((n-p)\times(n-p)),$$

$$\Phi_i^H\mathcal{A}\,\Phi_i = \Lambda_i \qquad i = 1,2. \tag{2.2.6}$$

(The superscript (p) at $\Lambda_{1/2}$, $\Phi_{1/2}$ will be dropped if possible without confusion.)
To have an even simple notation for the q smallest eigenvalues $\lambda_n,\dots,\lambda_{n-q+1}$ together with their eigenvectors, we use according to [PARLETT 1] a numeration with negative indices. The eigenvalues of $\{\mathcal{A},\mathcal{B}\}$ are further labelled as

$$\lambda_{-1} \le \lambda_{-2} \le \dots \le \lambda_{-n}, \tag{2.2.7}$$

thus,

$$\lambda_{-1} = \lambda_n \text{ and } \lambda_1 = \lambda_{-n}.$$

Analogously, we define φ_{-1} to be the eigenvector coressponding to λ_{-1} and

$$\Phi_{-1} = \Phi_{-1}^{(q)} = (\varphi_{-1},\dots,\varphi_{-q}) \in \mathbb{H}^q,$$

$$\Phi_{-2} = \Phi_{-2}^{(q)} = (\varphi_{-q-1},\dots,\varphi_{-n}) \in \mathbb{H}^{n-q},$$

$$\Lambda_{-1} = \Lambda_{-1}^{(q)} = \text{diag}\,(\lambda_{-1}, \ldots, \lambda_{-q}) \in \mathbb{R}^{q\times q},$$

$$\Lambda_{-2} = \Lambda_{-2}^{(q)} = \text{diag}\,(\lambda_{-q-1}, \ldots, \lambda_{-n}).$$

Summarizing these definitions, we state

$$\mathcal{A}\,\Phi_i = \mathcal{B}\,\Phi_i\,\Lambda_i, \qquad i = \pm 1,\ \pm 2,$$

$$\Phi_i^H\,\mathcal{B}\,\Phi_i = I, \qquad i = \pm 1,\ \pm 2,$$

$$\Phi_i^H\,\mathcal{A}\,\Phi_i = \Lambda_i, \qquad i = \pm 1,\ \pm 2,$$

$$\Phi_i^H\,\mathcal{A}\,\Phi_j = \Phi_i^H\,\mathcal{B}\,\Phi_j = 0, \qquad i \neq j \ (ij > 0).$$

The n-tuple of all eigenvectors is denoted by

$$\Phi = (\Phi_1 \vdots \Phi_2) = (\Phi_{-2} \vdots \Phi_{-1})$$

respectively, $\mathcal{A}\Phi = \mathcal{B}\Phi\Lambda$. (2.2.8)

Here,

$$\Lambda = \Phi^H \mathcal{A} \Phi \tag{2.2.9}$$

is the (nxn)-diagonal matrix containing all eigenvalues.
Note, that for any element $X \in \mathbb{H}^p$ the matrix

$$V = \Phi^H \mathcal{B}\, X \tag{2.2.10}$$

contains the coefficients of the eigenexpansions of the columns of X, and

$$X = \Phi V \ . \tag{2.2.11}$$

2.3. The Ritz-Projection of the Eigenvalue Problem $\{\mathcal{A}, \mathcal{B}\}$

Any p-dimensional subspace $\mathcal{U}$ in $\mathbb{H}$ is entirely described by p basis elements $y_1, \ldots, y_p$. We denote by

$$\mathcal{U} = \text{span}\ Y = \{ y = Yt : \forall t \in \mathbb{R}^p \}$$

the subspace spanned by the columns of $Y \in \mathbb{H}^p$. This is a p-dimensional subspace, if Y contains linearly independent columns. Let $\mathbb{F}_p$ be the set of all such elements in $\mathbb{H}^p$:

$$\mathbb{F}_p = \{ Y \in \mathbb{H}^p : \dim\ \text{span}\ Y = p \}$$

$$= \{ Y \in \mathbb{H}^p : Yt = 0 \Leftrightarrow t = 0 \}$$

For any $Y \in \mathbb{F}_p$, the Ritz-projection of the eigenvalue problem $\{\mathcal{A}, \mathcal{B}\}$ into the subspace span Y is defined to be the eigenvalue problem $\{\mathcal{P}_Y \mathcal{A} \mathcal{P}_Y, \mathcal{P}_Y \mathcal{B} \mathcal{P}_Y\}$ with the orthogonal projection operator $\mathcal{P}_Y$ onto span Y (see [SAAD 1, MEYER 3]). The matrix representation of the operator $\mathcal{P}_Y \mathcal{A} \mathcal{P}_Y$ in the basis Y is the (pxp)-matrix $(Y^H Y)^{-1}(Y^H \mathcal{A} Y)$, thus the projected eigenvalue problem is represented by the equivalent (pxp)-matrix eigenvalue problem

$$\{ Y^H \mathcal{A} Y, Y^H \mathcal{B} Y \} \qquad (2.3.1)$$

With the help of the Ritz-projection we choose a special basis X of the subspace span Y, the so-called Ritz-basis (containing Ritz-vectors). Let

$$A = Y^H \mathcal{A} Y,$$

$$B = Y^H \mathcal{B} Y,$$

then solve the full eigenproblem for the matrix pencil $\{A,B\}$, that is

$$Aq_i = \sigma_i B q_i, \qquad i = 1,\dots,p. \qquad (2.3.2)$$

With the diagonal matrix of the eigenvalues

$$\Delta = \operatorname{diag}(\sigma_1,\dots,\sigma_p)$$

and the eigenvector matrix $Q = (q_1,\dots,q_p)$, (2.3.2) is again the compact matrix equation

$$AQ = BQ\Delta . \qquad (2.3.3)$$

The element

$$X = YQ \in \mathbb{H}^p$$

is called Ritz-basis of the considered subspace span Y. With the obvious normalization

$$Q^T B Q = I$$

(the matrix B is also positive definite) the identities

$$X^H \mathcal{B} X = Q^T Y^H \mathcal{B} Y Q = I,$$
$$X^H \mathcal{A} X = \Delta \qquad (2.3.4)$$

follow from the definitions above.

Some important lemmas in connection with Ritz's process should be added here:

Lemma 2.3.1 [MEYER 1]. The eigenvalues in Δ depend on the given subspace span Y only, but do not depend on the specially chosen basis Y. Also one column x_i of the Ritz-basis does, if the eigenvalue δ_i is simple.

Lemma 2.3.2 [GANTMACHER 1]. If the eigenvalues δ_i are labelled analogously to (2.2.2), that is

$$\delta_1 \geq \delta_2 \geq \ldots \geq \delta_p \tag{2.3.5}$$

then the following inequalities hold

$$\delta_i \leq \lambda_i, \qquad i = 1,\ldots,p, \tag{2.3.6}$$

$$\delta_{p-i} \geq \lambda_{n-i}, \qquad i = 0,\ldots,p-1,$$

resp. $$\delta_{-i} \geq \lambda_{-i}, \qquad i = 1,\ldots,p. \tag{2.3.7}$$

A variety of further results connected with Ritz's process can be found in [PARLETT 1], see also the next chapter.

2.4. The Functional $\mu(X)$

The well-known properties of Rayleigh's quotient

$$\mu(x) = \frac{\langle \mathcal{A}x, x\rangle}{\langle \mathcal{B}x, x\rangle}$$

have been used to construct algorithms for the calculation of extremal eigenvalues (λ_1 or λ_n) of $\{\mathcal{A}, \mathcal{B}\}$, for example in [FADDEYEV/FADDEYEVA 1]. To compute next eigenvalues ($\lambda_2, \ldots, \lambda_p$ say), we also could maximize $\mu(x)$ with respect to orthogonalizations to previously calculated eigenvectors($\mathcal{B}$-inner product!). But, if for modern computers the calculation using more vectors (for example the execution of p matrix-vector-products $\mathcal{A}x_1, \ldots, \mathcal{A}x_p$ simultaneously) is as cheap as for one single vector, then the simultaneous calculation of p eigenvalues with corresponding eigenvectors becomes far more efficient. This especially occurs in connection with data-transfer problems for the large matrices or using vector processors. This leads to a generalization of Rayleigh's quotient belonging to p-dimensional subspaces span X, $X \in \mathbb{H}^p$.

Definition. For $X \in \mathbb{F}_p$ let

$$F(X) = (X^H \mathcal{B} X)^{-1}(X^H \mathcal{A} X)$$

and

$$\mu(X) = \operatorname{tr} F(X). \tag{2.4.1}$$

The following important properties are direct consequences of this definition:

(1) The matrix F has p real eigenvalues $\sigma_1, \dots, \sigma_p$. These are the Ritz-eigenvalues belonging to the subspace span X (compare 2.3.)

(2) From (1) together with Lemma 2.1.2 it follows:
The value $\mu(X)$ does not change under basis-transformations over the subspace span X, that is

$$\mu(X) = \mu(XT) \quad \text{for all regular } T \in \mathbb{H}^{p\times p}.$$

Thus, $\mu(X)$ does not depend on the <u>element</u> $X \in \mathbb{H}^p$, but on the <u>subspace</u> span X.

(3) From (1) and Lemma 2.3.2 we conclude

$$\mu(X) = \sum_{i=1}^{p} \sigma_i < \sum_{i=1}^{p} \lambda_i = \mathrm{tr}\Lambda_1^{(p)} = \mu(\Phi_1^{(p)}) \tag{2.4.2}$$

and respectively

$$\mu(X) > \sum_{i=1}^{p} \lambda_{-i} = \mathrm{tr}\Lambda_{-1}^{(p)} = \mu(\Phi_{-1}^{(p)}).$$

Thus,

$$\max_{X \in F_p} \mu(X) = \mu(\Phi_1) = \mathrm{tr}\Lambda_1$$

resp. $$\min_{X \in F_p} \mu(X) = \mu(\Phi_{-1}) = \mathrm{tr}\Lambda_{-1} \quad .$$

More precisely: The maximum of $\mu(X)$ is attained on a subspace spanned by the wanted eigenvectors $\varphi_1, \dots, \varphi_p$. Lemma 2.4.1 shows that this subspace is unique.

(4) As a consequence of the definition the functional $\mu(X)$ contains a collection of some simple nonlinear mappings and is thus differentiable.The derivatives directly follow from the chains rule of calculus for nonlinear functions (see [DÖHLER 1,3]).
That is, we have Taylor's rule

$$\mu(X+S) = \mu(X) + \mu'(X;S) + \tfrac{1}{2}\mu''(X;S,S) + \dots \tag{2.4.3}$$

with the first line derivative

$$\begin{aligned} \mu'(X;S) &= \mathrm{tr}\, F'(X;S) \\ &= 2\mathrm{tr}\, S^H R(X)(X^H \mathcal{B} X)^{-1} \\ &= \langle \nabla\mu(X), S \rangle \quad \text{(inner product in } \mathbb{H}^p\text{)}. \end{aligned} \tag{2.4.4}$$

Here, $F'(X;S)$ denotes the Fréchet derivative of $F(X)$ in the direction S,

$$F'(X;S) = (X^H \mathcal{B} X)^{-1}(S^H \mathcal{A} X + X^H \mathcal{A} S - (S^H \mathcal{B} X + X^H \mathcal{B} S)F(X)).$$

For the gradient direction $\nabla\mu(X)$ (2.4.4) yields

$$\nabla\mu(X) = 2\, R(X)(X^H \mathcal{B} X)^{-1},$$

with (2.4.5)

$$R(X) = \mathcal{A} X - \mathcal{B} X\, F(X)$$

the eigenvalue residual of X.
The second line derivative for two directions S_1 and S_2 results as

$$\mu''(X;S_1,S_2) = \mu''(X;S_2,S_1) = \mathrm{tr}\Big[2(S_2^H \mathcal{A} S_1 - S_2^H \mathcal{B} S_1 F(X)) -$$

$$- (X^H \mathcal{B} S_1 + S_1^H \mathcal{B} X)F'(X;S_2) - \qquad (2.4.6)$$

$$- (X^H \mathcal{B} S_2 + S_2^H \mathcal{B} X)F'(X;S_1)\Big](X^H \mathcal{B} X)^{-1}$$

<u>Lemma 2.4.1.</u> Let $\lambda_p > \lambda_{p+1}$, then the subspace span X with $\mu(X) = \mathrm{tr}\Lambda_1$ is unique,

$$\text{span } X = \text{span } \Phi_1.$$

<u>Proof.</u> Let $X \in \mathbf{F}_p$, then without loss of generality we choose the basis

$$X = \Phi_1 + \Phi_2 E$$

for a special $(n-p)\times p$-matrix E (compare 2.2.11). The assumption $\mu(X) = \mathrm{tr}\Lambda_1$ will lead to $E = \mathbf{0}$ due to the following calculation:

$$X^H \mathcal{B} X = I + E^T E,$$

$$X^H \mathcal{A} X = \Lambda_1 + E^T \Lambda_2 E,$$

$$0 = \mu(X) - \mathrm{tr}\Lambda_1 = \mathrm{tr}\,(I+E^TE)^{-1}(E^T\Lambda_2 E - E^T E \Lambda_1)$$

or

$$\mathrm{tr}\,(I+E^TE)^{-1}E^T\Lambda_2 E = \mathrm{tr}\,(I+E^TE)^{-1}E^TE\,\Lambda_1.$$

The left hand side is increased using Lemma 2.1.3

$$\mathrm{tr}\,(I+E^TE)^{-1}E^T\Lambda_2 E \leq \lambda_{p+1}\ \mathrm{tr}\,(I+E^TE)^{-1}E^TE.$$

For the right hand side note that $(I+E^TE)^{-1}E^TE$ is a symmetric matrix (a function of E^TE), thus again,

$$\operatorname{tr}(I+E^TE)^{-1}E^TE\,\Lambda_1 \geq \lambda_p \operatorname{tr}(I+E^TE)^{-1}E^TE.$$

Hence, $E \neq 0$ yields $\lambda_p \leq \lambda_{p+1}$. ∎

Next, we consider the problem of choosing this p-dimensional subspace (say span X) as a part of a higher dimensional subspace span Y ($Y \in \mathbb{H}^{p+m}$, $m \geq 1$) by maximizing $\mu(X)$.

Assume $Y \in \mathbb{F}_{p+m}$ (i.e. span Y has full dimension p+m), then any rectangular matrix $U \in \mathbb{R}^{(p+m)xp}$ defines an element $\tilde{X} = YU \in \mathbb{H}^p$, respectively a p-dimensional subspace

$$\operatorname{span} \tilde{X} \subset \operatorname{span} Y,$$

if U has full rank p.

Here, we answer the question: What matrix U leads to the subspace span X with maximal value $\mu(X)$? The answer is given with the help of the Ritz-projection of $\{\mathcal{A},\mathcal{B}\}$ into the subspace span Y, that is, we define the symmetric (p+m)x(p+m)-matrices

$$A = Y^H \mathcal{A}\, Y,$$

$$B = Y^H \mathcal{B}\, Y$$

and solve the full eigenproblem for the matrix pencil $\{A,B\}$. Thus

$$\delta_1 \geq \delta_2 \geq \ldots \geq \delta_p \geq \delta_{p+1} \geq \ldots \geq \delta_{p+m} \tag{2.4.7}$$

denote the eigenvalues with

$$q_1,\ q_2,\ \ldots,\ q_p,\ q_{p+1},\ \ldots,\ q_{p+m} \in \mathbb{R}^{p+m}$$

the corresponding eigenvectors. Let $Q \in \mathbb{R}^{(p+m)xp}$ be the matrix, having the columns $q_1,\ldots,q_p$ of the first p eigenvectors belonging to the p largest eigenvalues of $\{A,B\}$.

According to the following lemma the matrix Q is that matrix maximizing $\mu(YU)$ over U.

<u>Lemma 2.4.2.</u> With the above definitions the inequality

$$\mu(\tilde{X}) = \mu(YU) \leq \mu(X) = \mu(YQ) = \sum_{i=1}^{p} \delta_i$$

holds for all $U \in \mathbb{R}^{(m+p)xp}$ of full rank p.

<u>Proof.</u> With

$$\tilde{A} = \tilde{X}^H \mathcal{A}\, \tilde{X}, \qquad \tilde{B} = \tilde{X}^H \mathcal{B}\, \tilde{X}$$

we have

$$\mu(\tilde{X}) = \operatorname{tr} \tilde{B}^{-1}\tilde{A} = \sum_{i=1}^{p} \tilde{\sigma}_i,$$

where $\tilde{\sigma}_i$ are the eigenvalues of the pencil $\{\tilde{A},\tilde{B}\}$.
But

$$\tilde{A} = (YU)^H \mathcal{A}\, YU = U^T AU,$$

$$\tilde{B} = (YU)^H \mathcal{B}\, YU = U^T BU,$$

thus by the Ritz-argument (according to Lemma 2.3.2)

$$\tilde{\sigma}_i \leq \sigma_i, \qquad i = 1,\dots,p,$$

which completes the proof. ∎

Corollary. The problem of finding the "best" p-dimensional subspace span X (with maximal $\mu(X)$) in a higher dimensional subspace is uniquely solved if

$$\sigma_p > \sigma_{p+1}$$

The constructions before Lemma 2.4.2 form a Ritz-basis X of the resulting subspace, thus according to (2.3.4) the element $X = YQ$ fulfils the following special formulas

$$X^H \mathcal{B}\, X = I, \tag{2.4.8}$$

$$X^H \mathcal{A}\, X = F(X) = \operatorname{diag}(\sigma_1,\dots,\sigma_p) = \Delta, \tag{2.4.9}$$

$$R(X) = \mathcal{A} X - \mathcal{B} X \Delta = (r_1,\dots,r_p) \tag{2.4.10}$$

with

$$r_j = \mathcal{A} x_j - \sigma_j \mathcal{B} x_j, \qquad j = 1,\dots,p.$$

The simplification in the residual formula (2.4.10) (compared with (2.4.5) for arbitrary $X \in \mathbb{H}^p$) is also a marked representation of a Ritz-basis and leads to the following lemma.

Lemma 2.4.3. The columns of the eigenvalue residual R(X) for the Ritz-basis $X = YQ$ ($X \in \mathbb{H}^p, Y \in \mathbb{H}^{p+m}$) are orthogonal to all columns of Y,

$$Y^H R(X) = \mathbb{O} \tag{2.4.11}$$

(resp. the subspaces span R(X) and the given span Y are orthogonal).

Proof. A direct calculation of

$$Y^H R(X) = Y^H \mathcal{A} YQ - Y^H \mathcal{B} YQ \Delta = AQ - BQ \Delta = \mathcal{O}$$

shows the influence of the eigenvalue identity for the pencil $\{A,B\}$, that is the matrix equation

$$AQ = BQ \Delta$$

defining the matrices Δ and Q of eigenvalues and -vectors, respectively.

Remark. This lemma is of some practical importance for $m > 0$ only. In this case the orthogonality $Y^H R(X) = \mathcal{O}$ also means a special representation of X containing Ritz-vectors. For $m = 0$ (i.e. span X = span Y) the relation $Y^H R(X) = \mathcal{O}$ would be true for any basis X of that subspace not for Ritz-vectors only. The latter follows from the definition of R(X) in (2.4.5):

1. $Y^H R(Y) = \mathcal{O} \qquad \forall Y \in \mathbb{F}_p,$ (2.4.12)

2. $R(YU) = R(Y)U$ (2.4.13)

   for all regular (pxp)-matrices U.

Thus any basis $X = YU$ of the subspace span Y fulfils

$$Y^H R(X) = Y^H R(YU) = Y^H R(Y)U = \mathcal{O}. \quad (2.4.14)$$

For more general applications another functional depending on two subspaces is defined now (compare [CULLUM 1]). By maximizing $\mu(X)$ over $X \in \mathbb{F}_p$ we are able to calculate span Φ_1 and the p largest eigenvalues, by minimizing $\mu(X)$ (or maximizing $-\mu(X)$), analogously, smallest eigenvalues of $\{\mathcal{A}, \mathcal{B}\}$ and span Φ_{-1} are obtained. With the definition

$$\mu(X_1,X_2) = \mu(X_1) - \mu(X_2) \quad \forall X_1 \in \mathbb{F}_p,\ X_2 \in \mathbb{F}_q \quad (2.4.15)$$

we are able to calculate simultaneously the p largest eigenvalues $\lambda_1, \dots, \lambda_p$ and the q smallest $\lambda_{-1}, \dots, \lambda_{-q}$ together with their eigenvectors by maximizing $\mu(X_1,X_2)$ over all pairs $X_1 \in \mathbb{F}_p$ and $X_2 \in \mathbb{F}_q$. With $p = 0$ or $q = 0$ the above is contained as a special case.
(The same letter μ for this new functional will not confuse the reader, because the new one is distinguished from the former one by its two arguments, which always will be written.)
The properties of $\mu(X_1,X_2)$ are direct consequences of the properties obtained for $\mu(X)$:

(1) $\mu(X_1,X_2)$ is constant for all bases X_1T_1 of span X_1 and X_2T_2 of span X_2 (T_1 square regular matrices of proper dimensions), unfortunately $\mu(X_1,X_2)$ is not constant over the subspace span $(X_1 \vdots X_2)$, the direct sum of span X_1 and span X_2. Thus, exactly $\mu(X_1,X_2)$ maps two subspaces into the real numbers.

(2) $\mu(X_1,X_2) \leqslant \mu(\phi_1^{(p)}) - \mu(\phi_{-1}^{(q)})$ (2.4.16)

for all $X_1 \in \mathbb{F}_p$, $X_2 \in \mathbb{F}_q$

and the maximum of $\mu(X_1,X_2)$ is attained at

span X_1 = span $\phi_1^{(p)}$,

span X_2 = span $\phi_{-1}^{(q)}$.

(3) For the gradient it follows

$$\nabla\mu(X_1,X_2) = 2\left(R(X_1)(X_1^H \mathcal{B} X_1)^{-1} \;\vdots\; - R(X_2)(X_2^H \mathcal{B} X_2)^{-1}\right)$$

$$= (\nabla\mu(X_1) \;\vdots\; -\nabla\mu(X_2)) \in \mathbb{H}^{p+q}. \qquad (2.4.17)$$

(4) Let $Y \in \mathbb{H}^{p+q+m}$ with $m > 0$ and linearly independent columns, then the "best" pair of a p-dimensional subspace span $X_1 \subset$ span Y and a q-dimensional subspace span $X_2 \subset$ span Y, for which $\mu(X_1,X_2)$ is maximized, is constructed analogously to Lemma 2.4.2:

For $A = Y^H \mathcal{A} Y \in \mathbb{R}^{(p+q+m)x(p+q+m)}$

and $B = Y^H \mathcal{B} Y$

we solve the full eigenproblem of the pencil $\{A,B\}$:

$$Aq_i = \delta_i B q_i, \qquad i = 1,\dots,(p+q+m)$$

$$\text{or} \quad i = -1,\dots,-(p+q+m),$$

$$\delta_1 \geq \dots \geq \delta_{p+q+m}$$

resp.

$$\delta_{-1} \leqslant \dots \leqslant \delta_{-(p+q+m)}.$$

Then, if

Δ_1 = diag $(\delta_1,\dots,\delta_p)$, (p largest eigenvalues)

$Q_1 = (q_1,\dots,q_p) \in \mathbb{R}^{(p+q+m)xp}$,

and

Δ_{-1} = diag $(\delta_{-1},\dots,\delta_{-q})$, (q smallest eigenvalues)

$Q_{-1} = (q_{-1},\dots,q_{-q}) \in \mathbb{R}^{(p+q+m)xq}$,

the definitions

$X_1 = YQ_1$ and $X_2 = YQ_{-1}$

yield maximal value $\mu(X_1,X_2)$.

Lemma 2.4.4. With the above definitions

$$\mu(\tilde{X}_1,\tilde{X}_2) \leq \mu(X_1,X_2) = \mathrm{tr}\,\Delta_1 - \mathrm{tr}\,\Delta_{-1}$$

holds for all $\tilde{X}_1 = YU_1$ and $\tilde{X}_2 = YU_2$ with matrices $U_1 \in \mathbb{R}^{(p+q+m)xp}$ and $U_2 \in \mathbb{R}^{(p+q+m)xq}$ of full rank.

The proof is equivalent to Lemma 2.4.2.

Notes and References

1. Many authors discussed the functional $\mu(X)$, which is a basic tool for the construction of simultaneous Rayleigh-quotient-minimization methods, especially [KY FAN 1, CULLUM/ DONATH/ WOLFE 1, CULLUM 1, SAMEH/ WISNIEWSKI 1, GOLEBIEWSKI 1, LONGSINE/ McCORMICK 1] and others . Often the unsymmetric matrix $F(X)$ is avoided using the constraint $X^H \mathcal{B} X = I$.

On the other hand, for some theoretical investigations this is by no means an advantage. For example, using the constraint

$$X^H \mathcal{B} X = I \qquad (2.4.18)$$

we are going to investigate the behaviour of $\mu(X+S)$ for all $S \in \mathbb{H}^p$ with

$$S^H \mathcal{B} X = \mathbb{O} \quad . \qquad (2.4.19)$$

Here, the constraint

$$(X+S)^H \mathcal{B} (X+S) = I \qquad (2.4.20)$$

makes no sense, because all three formulas (2.4.18) - (2.4.20) yield

$$S = 0.$$

Hence, there is the only chance to use

$$\mu(X+S) = \mathrm{tr}\, F(X+S)$$

without any constraints, if (2.4.18) and (2.4.19) are fulfilled, which embodies a real appropriate case in proofs of convergence at Chapter 4.

2. The calculus formulas for $\mu(X)$ in property (4) are only used to construct the direction of steepest descent or ascent $\nabla\mu(X)$, all other theorems are purely algebraical. The second derivative is necessary to generalize the ideas of conjugate directions to the functional $\mu(X)$ in a full simultaneous way due to [DÖHLER 1,3].

3. SIMULTANEOUS ALGORITHMS FOR MAXIMIZING $\mu(X)$

This chapter presents the basic algorithmic idea, which leads to some variants in choosing special directions of movement. Here, a more complicate choice consists in the generalization of the conjugate-gradient method to the functional $\mu(X)$. The theoretical foundation for that method will be the topic of the second part of this chapter.

3.1. The Basic Algorithm

In this chapter, we state the basic algorithm according to the following idea. Let $X = X_k \in \mathbb{H}^p$ be the k-th iterated element. For purposes of brevity we drop the iteration index k, thus we denote the next iterated element X_{k+1} by $\hat{X}$ and also all other quantities of the next step of iteration will be correspondingly designated in this way. Respectively, a quantity of the preceding iteration step, say X_{k-1}, is denoted by $\check{X}$. Throughout this paper, $S \in \mathbb{H}^m$ will be any "direction of movement", depending on X or only depending on the step number of the iteration process. Then $\hat{X} \in \mathbb{H}^p$ is an element with

$$\text{span } \hat{X} \subset \text{span } (X \vdots S)$$

and maximal value $\mu(\hat{X})$.
Following Chapter 2 the construction of $\hat{X}$ using Ritz's process leads to our

Basic Algorithm

(1) construct S

(2) $$A := (X \vdots S)^H \mathcal{A} (X \vdots S) \tag{3.1.1}$$

$$A = \begin{pmatrix} A_{11} & A_{12} \\ A_{12}^T & A_{22} \end{pmatrix}$$

with $A_{11} = X^H \mathcal{A} X$,
$A_{12} = X^H \mathcal{A} S$,
$A_{22} = S^H \mathcal{A} S$,

$$B := (X \vdots S)^H \mathcal{B} (X \vdots S) \tag{3.1.2}$$
$$= \begin{pmatrix} B_{11} & B_{12} \\ B_{12}^T & B_{22} \end{pmatrix}$$

with $B_{11} = X^H \mathcal{B} X$,
$B_{12} = X^H \mathcal{B} S$,
$B_{22} = S^H \mathcal{B} S$.

(3) Solve the generalized matrix eigenvalue problem $\{A,B\}$, calculate p largest eigenvalues $\hat{\sigma}_1 \geq \dots \geq \hat{\sigma}_p$ with corresponding eigenvectors $q_1,\dots,q_p$, that is

$$AQ = BQ\hat{\Delta} \tag{3.1.3}$$

with

$$\hat{\Delta} = \mathrm{diag}\,(\sigma_1,\dots,\sigma_p) \in \mathbb{R}^{p\times p},$$

$$Q = (q_1,\dots,q_p) \in \mathbb{R}^{(p+m)\times p}.$$

Let

$$Q^T BQ = I.$$

(4) $$\hat{X} := (X \vdots S)Q.$$

<u>Corollaries</u>

1. $\mu(\hat{X}) = \mathrm{tr}\hat{\Delta}$,
   furthermore the approximate eigenvalue $\hat{\sigma}_i$ belongs to the column $\hat{x}_i$ $(i=1,\dots,p)$.
2. (3.1.3) leads to
   $\hat{B}_{11} = \hat{X}^H\mathcal{B}\hat{X} = I, \quad \hat{A}_{11} = \hat{X}^H\mathcal{A}\hat{X} = \hat{\Delta}$
   (compare with 2.3), hence these two properties are fulfilled for all iterates X (may be except for the starting element X_o) that is,
   $$B_{11} = X^H\mathcal{B}X = I,$$
   $$A_{11} = X^H\mathcal{A}X = \Delta = \mathrm{diag}\,(\sigma_1,\dots,\sigma_p). \tag{3.1.4}$$
   Here, σ_i denote the approximate eigenvalues belonging to x_i and $\sigma_1 \geq \dots \geq \sigma_p$.
3. The two (pxp)-leading submatrices of A and B in (3.1.1)/(3.1.2) are known to be Δ and I resp.
   Thus, by the Ritz-argument,
   $$\hat{\sigma}_i \geq \sigma_i, \qquad i = 1,\dots,p,$$
   and
   $$\mu(\hat{X}) = \mathrm{tr}\hat{\Delta} \geq \mathrm{tr}\Delta = \mu(X).$$
4. The feasibility of that step of the iteration is guaranteed, if dim span $(X \vdots S) = p+m$. From (3.1.4) we have dim span $(X \vdots S) \geq p$. The number m need not be fixed. We can drop some columns of S, yielding up dim span $(X \vdots S) = p+m$ with a smaller number $m \geq 1$. Thus this algorithm is always feasible, if there exists at least one column of S, which is linearly independent of the columns of X.

Paying attention to some special choices of S, we can distinguish some variants of this basic algorithm. If S contains some coordinate unit vectors (cyclically chosen over the iteration steps), we obtain the simultaneous group coordinate relaxation method as investigated by [SCHWARZ 1, GOSE 1, HANZAL 1, SKURT 1,2]. The rate of convergence of this method is known to be very poor and up to now there are no possibilities of preconditioning ideas to speed up this method. Especially these preconditionings make another method very interesting and useful, the method of steepest ascent (gradient method).
This consists in the choice

$$S := R(X) = \mathcal{A}X - \mathcal{B}X\Delta. \qquad (3.1.5)$$

For this method we show later (Chapter 5) that with an appropriate definition of $\mathcal{A}$ and $\mathcal{B}$ (together with the inner product) a preconditioned version of the simple gradient method arises. The basic idea of this method (for symmetric matrices $\mathcal{A}, \mathcal{B}$ and Euclidian inner product without preconditionings) was investigated by [SCHWARZ 1,3; SAMEH/WISNIEWSKI 1; LONGSINE/McCORMICK 1]. Some authors tried to generalize the ideas of the conjugate gradient iteration method to the functional $\mu(X)$ for further acceleration of the method of steepest ascent. [GERADIN 1] and [FRIED 1] gave a foundation of the method for the Rayleigh-quotient functional $\mu(x) = \langle \mathcal{A}x, x\rangle / \langle \mathcal{B}x, x\rangle$. Using similar formulas (especially $S = R + \beta\check{S}$) for a simultaneous maximizing of $\mu(X)$, only a slight acceleration of the gradient method is obtained (compare [CULLUM 1]). The more effective formula

$$S = R + \check{S}T$$

with a (pxp)-diagonal matrix T is due to [LONGSINE/McCORMICK 1]. Here, T is calculated by considering the single columns of S, $\check{S}$ and R

$$s_i = r_i + \beta_i \check{s}_i, \qquad i = 1, \ldots, p.$$

The full generalization of GERADIN's and FRIED's algorithms to the simultaneous functional was presented by [DÖHLER 1,2,3]. Here, the direction of movement

$$S = R + WT$$

is constructed with a dense (pxp)-matrix T and the special element $W \in \mathbb{H}^p$ (instead of $\check{S}$), for which

$$\operatorname{span}(X \vdots W) = \operatorname{span}(\check{X} \vdots \check{S})$$

together with some orthogonality conditions hold. These facts ensure that T becomes easy comparable based on a near-maximizing argument of $\mu(X+S)$. The details of this algorithm are the topic of the next chapter.

3.2. The Simultaneous Conjugate Gradient Algorithm for $\mu(X)$

The main problem to state this algorithm as a generalization of [FRIED's 1] and [GERADIN's 1] ideas to the more complicated functional $\mu(X)$ consists in the foundation of the (pxp)-matrix T. [DÖHLER 1,2,3] found a slight change in the definition, which allows an easy computation of T. In this algorithm any step of the iterations starts with $X \in \mathbb{H}^p$ together with an additional element $W \in \mathbb{H}^p$ for which the identities

$$\text{span}\,(X \vdots W) = \text{span}\,(\check{X} \vdots \check{S}), \tag{3.2.1}$$

$$(X \vdots W)^H \mathcal{B}\,(X \vdots W) = I, \tag{3.2.2}$$

$$(X \vdots W)^H \mathcal{A}\,(X \vdots W) = \Delta = \text{diag}\,(\Delta_1, \Delta_2) \tag{3.2.3}$$

hold. Here, we denote by

$$\Delta_1 = \text{diag}\,(\sigma_1, \dots, \sigma_p),$$

$$\sigma_1 \geq \dots \geq \sigma_p,$$

the approximations of the eigenvalues belonging to X. Furthermore $\Delta_2 = \text{diag}\,(\sigma_{p+1}, \dots, \sigma_{2p})$ is also calculated from a (2px2p)-eigenvalue problem (the remaining, smaller eigenvalues). The construction of $\hat{X}$ and $\hat{W}$ during the actual step of the iteration is shown in the following algorithm.

Conjugate Gradient Method for $\mu(X)$

(1) $R := \mathcal{A}X - \mathcal{B}X\,\Delta_1$.

(2) Calculate $T \in \mathbb{R}^{p\times p}$ using X, W, R.

(3) $S = R + WT$.

(4) $A = (X \vdots S)^H \mathcal{A}\,(X \vdots S)$,

$B = (X \vdots S)^H \mathcal{B}\,(X \vdots S)$.

Solve the eigenvalue problem of $\{A,B\}$: calculate all 2p eigenvalues

$$\hat{\sigma}_1 \geq \dots \geq \hat{\sigma}_I \geq \hat{\sigma}_{p+1} \geq \dots \geq \hat{\sigma}_{2p},$$

$$\hat{\Delta} = \text{diag}\,(\hat{\Delta}_1, \hat{\Delta}_2),$$

$$\hat{\Delta}_1 = \text{diag}\,(\hat{\sigma}_1, \dots, \hat{\sigma}_p),$$

$$\hat{\Delta}_2 = \text{diag}\,(\hat{\sigma}_{p+1}, \dots, \hat{\sigma}_{2p}).$$

Let $Q = (q_1, \dots, q_{2p})$ be the full matrix of eigenvectors

$$Aq_i = \hat{\sigma}_i B q_i, \quad i = 1, \dots, 2p,$$

$$Q^T B Q = I.$$

(5) $(\hat{X} \vdots \hat{W}) = (X \vdots S)Q$.

From the steps (4) and (5) we obtain $(\hat{X} \vdots \hat{W})$ as the Ritz-basis of the subspace span $(X \vdots S)$, thus all assumptions (3.2.1) to (3.2.3) are again fulfilled for $\hat{X}$ and $\hat{W}$.

It remains to state an appropriate formula for step (2), that is the construction of T. The functional $\mu(X)$ is a rational, but not a quadratic one (as in the case of linear systems), that is, the Taylor series of $\mu(X+S)$ is infinite. On the other hand, for small S we can approximate $\mu(X+S)$ by the first three terms of Taylor's series yielding a quadratic approximate functional $\mu_X(S)$

$$\mu_X(S) = \mu(X) + \mu'(X;S) + \frac{1}{2}\mu''(X;S,S).$$

According to the principles of [FRIED 1] or [GERADIN 1] for Rayleigh's quotient it becomes possible to construct "conjugate directions" S and $\check{S}$ with the help of maximizing $\mu_X(S) = \mu_X(R+WT)$ over T.

Note that for any differentiable functional $\nu(X)$ the maximum $\nu(R+\alpha W)$ over $\alpha \in \mathbb{R}$ could be calculated from $\nu'(R+\alpha W;W) = 0$, which is one condition defining the one number α. Here, the more complicate task, maximizing $\nu(R+WT)$ over all (pxp)-matrices T, leads to

$$\nu'(R+WT;WE) = 0 \quad \forall E \in \mathbb{R}^{p \times p},$$

which (possibly) can be transformed into a matrix equation defining T. Now let T be defined to maximize the functional $\mu_X(S) = \mu_X(R+WT)$. Then the first line derivative of $\mu_X(S)$ into a direction $S_1 \in \mathbb{R}^p$ is calculated from the identity

$$\mu_X(S+S_1) - \mu_X(S) = \mu'(X;S_1) + \mu''(X;S,S_1) + \frac{1}{2}\mu''(X;S_1,S_1)$$

to be

$$\mu_X'(S;S_1) = \mu'(X;S_1) + \mu''(X;S,S_1).$$

Thus, maximizing $\mu_X(S)$ over T leads to

$$\mu'_X(S;WE) = 0 \quad \forall E \in \mathbb{R}^{p \times p}$$

or

$$\mu'(X;WE) + \mu''(X;S,WE) = 0 \qquad \forall E \in \mathbb{R}^{p\times p}$$

with $S = R + WT$. Now (2.4.4) yields

$$\mu'(X;WE) = 2\mathrm{tr}\,(WE)^H R = 2\mathrm{tr}\,E^T W^H R = 0,$$

because of the orthogonality conditions (3.2.2) and (3.2.3).
Thus, the generalized "conjugate directions" S and W are defined in the sense that S and WE become "orthogonal" due to the second line derivative of $\mu(X)$ for all E. [1]

$$\mu''(X;S,WE) = 0 \qquad \forall E \in \mathbb{R}^{p\times p}. \tag{3.2.4}$$

The ingenious use of the special element W (instead of $\check{S}$) allows a solution of that equation defining T. From (2.4.6) and the assumptions (3.2.1) - (3.2.3) we obtain

$$\mu''(X;S,WE) = 2\mathrm{tr}(E^T W^H \mathcal{A}\, S - E^T W^H \mathcal{B}\, S\, \Delta_1),$$

while all other inner products vanish.
Hence, (3.2.4) corresponds to the matrix equation

$$W^H \mathcal{A}\, S - W^H \mathcal{B}\, S\, \Delta_1 = \mathcal{O},$$

which defines T from the Lyapunov equation

$$W^H \mathcal{A}\, R - W^H \mathcal{B}\, R\, \Delta_1 = T\,\Delta_1 - \Delta_2 T. \tag{3.2.5}$$

With the help of (3.2.5) any element t_{ij} of T is calculated by

$$t_{ij} = (\delta_j - \delta_{p+i})^{-1}\,(\langle \mathcal{A} r_j, w_i\rangle - \delta_j \langle \mathcal{B} r_j, w_i\rangle), \tag{3.2.6}$$

whenever $\delta_p > \delta_{p+1}$ holds.

---

1) If the above technique is used in connection with the quadratic functional

$$\mathrm{tr}(\tfrac{1}{2}X^H \mathcal{A}\, X - B^H X)$$

to solve p linear systems $\mathcal{A}X = B \in \mathbb{H}^p$ simultaneously, the block-conjugate gradient method due to [O'LEARY 1] arises. Thus, our choice of T represents an obvious generalization to nonquadratic functionals.

On the other hand, if $\sigma_p = \sigma_{p+1}$, the computation of t_{1p} becomes impossible. For this exceptional case [DÖHLER 1] suggests to drop some columns of W by using a rectangular T fulfilling (3.2.5). On the other hand, for a small difference $(\sigma_p - \sigma_{p+1})$ also very large numbers t_{1p} seem to be crucial from a numerical point of view.

From a continuity argument we change (3.2.6) to the following definition for avoiding too large numbers $|t_{ij}|$:

Let

$$\varepsilon_{ij} = \min\{\gamma, (\sigma_j - \sigma_{p+i})\}, \tag{3.2.7}$$

for a given small constant γ, then define T according to

$$t_{ij} = \frac{1}{\varepsilon_{ij}} \left(\langle \mathcal{A} r_j, w_i \rangle - \sigma_j \langle \mathcal{B} r_j, w_i \rangle \right). \tag{3.2.8}$$

Note that with this altered definition the elements t_{ij} are bounded, because $\varepsilon_{ij} \geq \gamma$ and the inner products are bounded from (3.2.2)/(3.2.3).

We state the proof of convergence to eigenelements using this boundness of T in Chapter 4. Thus, if γ is small enough and $\lambda_p > \lambda_{p+1}$, the choice $\varepsilon_{ij} = \gamma$ (for some i,j) will only occur in early iteration steps, because σ_j converge to λ_j (for j=1,...,p).

The proof of convergence to eigenelements will use the following lemma giving an estimate for the matrix $S^H \mathcal{B} S$.

<u>Lemma 3.2.1.</u> The matrices T and $S^H \mathcal{B} S$ are bounded by

$$\|T\|_2 \leq \frac{p}{\gamma} \|R^H \mathcal{B} R\|_2^{\frac{1}{2}} (\lambda_1 - \lambda_n), \tag{3.2.9}$$

$$\|S^H \mathcal{B} S\|_2 \leq \|R^H \mathcal{B} R\|_2 \left(1 + \frac{p}{\gamma}(\lambda_1 - \lambda_n)\right)^2. \tag{3.2.10}$$

<u>Proof.</u> We have for the elements of T

$$\begin{aligned} |t_{ij}| &\leq \frac{1}{\gamma} \left| \langle r_j, (\mathcal{A} - \sigma_j \mathcal{B}) w_i \rangle \right| = \\ &= \frac{1}{\gamma} \left| \langle \mathcal{B}^{\frac{1}{2}} r_j, \ \mathcal{B}^{-\frac{1}{2}} (\mathcal{A} - \sigma_j \mathcal{B}) w_i \rangle \right| \\ &\leq \frac{1}{\gamma} \ \langle \mathcal{B} r_j, r_j \rangle^{\frac{1}{2}} \ \left\| \mathcal{B}^{-\frac{1}{2}} (\mathcal{A} - \sigma_j \mathcal{B}) w_i \right\| . \end{aligned}$$

If

$$w_i = \Phi \, b_i, \qquad b_i \in \mathbb{R}^n,$$

denotes the decomposition of w_i with respect to the eigenbasis, then the second norm becomes

$$\|\mathcal{B}^{-\frac{1}{2}}(\mathcal{A}-\sigma_j\mathcal{B})\Phi b_i\| = \|\mathcal{B}^{\frac{1}{2}}\Phi(\Lambda-\sigma_j I)b_i\| \leq \lambda_1 - \lambda_n,$$

because of $1 = \langle \mathcal{B} w_i, w_i \rangle = b_i^T b_i$ and $\Phi^H \mathcal{B} \Phi = I$.

Furthermore, $\langle \mathcal{B} r_j, r_j \rangle$ is a main diagonal element of the matrix $R^H \mathcal{B} R$, thus

$$\|T\|_2 \leq p \cdot \max_{i,j} |t_{ij}| \leq \frac{p}{\gamma} \rho^{\frac{1}{2}}(R^H\mathcal{B} R)\cdot(\lambda_1-\lambda_n).$$

From the definition $S = R + WT$ the matrix $S^H\mathcal{B} S$ becomes

$$S^H\mathcal{B} S = R^H\mathcal{B} R + T^T W^H \mathcal{B} R + R^H \mathcal{B} WT + T^T T \qquad (3.2.11)$$

and it remains to estimate $W^H\mathcal{B} R$ (using Lemma 2.1.1):

$$\|W^H\mathcal{B} R\|_2 = \|(\mathcal{B}^{\frac{1}{2}}W)^H(\mathcal{B}^{\frac{1}{2}}R)\|_2 \leq \|W^H\mathcal{B} W\|_2^{\frac{1}{2}} \|R^H\mathcal{B} R\|_2^{\frac{1}{2}} = \|R^H\mathcal{B} R\|_2^{\frac{1}{2}}$$

The norm inequality in (3.2.11) completes the proof. ∎

Some notes to the implementation and the complexity of this algorithm shall be added here. At first note that one step of the iteration requires the calculation of the elements

$\mathcal{A}X$, $\mathcal{B}X$ (to calculate R) and

$\mathcal{A}R$, $\mathcal{B}R$.

Furthermore, 4 non-symmetric and two symmetric (pxp)-matrices containing $\approx 6p^2$ inner products are necessary:
The two matrices $W^H\mathcal{A} R$ and $W^H\mathcal{B} R$ for defining T will be also used for the calculation of the Ritz-matrices A and B to avoid the further multiplications $\mathcal{A}S$ and $\mathcal{B}S$.
This follows from the block structure of A and B,

$$A = \begin{pmatrix} A_{11} & A_{12} \\ A_{21} & A_{22} \end{pmatrix}, \qquad A_{ij} \in \mathbb{R}^{p\times p}, \text{ resp. } B_{ij},$$

and from the orthogonalities (3.2.2)/(3.2.3) yielding

$$A_{11} = \Delta_1,$$

$$A_{12} = A_{21}^T = X^H\mathcal{A} S = X^H\mathcal{A} R,$$

$$A_{22} = S^H\mathcal{A} S = R^H\mathcal{A} R + T^T(W^H\mathcal{A} R) + (W^H\mathcal{A} R)^T T + T^T\Delta_2 T,$$

$$B_{11} = I,$$

$$B_{12} = B_{21}^T = X^H \mathcal{B} S = X^H \mathcal{B} R,$$

$$B_{22} = S^H \mathcal{B} S = R^H \mathcal{B} R + T^T(W^H \mathcal{B} R) + (W^H \mathcal{B} R)^T T + T^T T.$$

Hence, in addition to $W^H \mathcal{A} R$ and $W^H \mathcal{B} R$ the matrices $X^H \mathcal{A} R$, $X^H \mathcal{B} R$ and the symmetric matrices $R^H \mathcal{A} R$ and $R^H \mathcal{B} R$ are to calculate. With this description the more simple algorithm of steepest ascent (simultaneous gradient method) arises as a special case for $T = \mathcal{O}$, where the complexity decreases to two non-symmetric and two symmetric matrices of inner products, furthermore the calculation of $\hat{W}$ in step (5) vanishes.

3.3. Another Simultaneous Algorithm, Which Uses Information from the Previous Iteration Step

The main iteration formula of the simultaneous conjugate gradient method of Chapter 3.2. can be written in the following form

$$\hat{X} = (X \vdots R + WT)Q$$

and with

$$Q = \begin{pmatrix} Q_1 \\ Q_2 \end{pmatrix}, \qquad Q_{1/2} \in \mathbb{R}^{p \times p},$$

this leads to

$$\hat{X} = XQ_1 + RQ_2 + WTQ_2. \tag{3.3.1}$$

According to 3.2, T was chosen from a near-maximizing argument of $\mu(\hat{X})$. If the matrix TQ_2 is replaced by a third matrix Q_3 independent of $Q_{1/2}$, an even larger value $\mu(\hat{X})$ can be obtained by maximizing over Q_1, Q_2, and Q_3. That is, $\hat{X}$ gets largest functional value $\mu(\hat{X})$ over the 3p-dimensional subspace

$$\text{span } (X \vdots R \vdots W)$$

for fixed p-tuples X, R and W of the actual step of iteration. For such an algorithm a new problem occurs, which consists in the definition of $\hat{W}$. For the simultaneous conjugate gradient method span $\hat{W}$ was constructed to be the orthogonal complement of span $\hat{X}$ in span $(X \vdots S)$ = span $(\hat{X} \vdots \hat{W})$. Here, such an approach is impossible, because of the dimension 3p of the actual subspace span $(X \vdots R \vdots W)$ for defining $\hat{X}$ and $\hat{W}$. On the other hand, W was a substitute for $\check{S}$ in the simultaneous conjugate gradient method, thus, a convenient choice of $\hat{W}$ will be the "direction of movement" generated at the actual step of iteration, that is

$$\hat{W} = RQ_2 + WQ_3. \tag{3.3.2}$$

Note that $\hat{W} = (RQ_2 + WQ_3)U$ for any regular (pxp)-matrix U does not imply any change of the resulting algorithm, because for the next step of iteration

$$\text{span } (\hat{X} \vdots \hat{R} \vdots \hat{W}) = \text{span } (\hat{X} \vdots \hat{R} \vdots \hat{W}U) \qquad \forall U$$

holds. Hence, the results of this next step do not depend on U.

This new algorithm would never be recommended, if the computational costs would exceed those of the simultaneous conjugate gradient method. The definition (3.3.2), however, decreases these costs substantially. On the other hand, the above derivation of this algorithm (especially the definition of W according to (3.3.2)) is a pure empiric result. The comparison of both algorithms in numerical experiments justifies this new construction. Both algorithms always need the same number of necessary iterations for a given problem (up to one or two, possibly). Thus, both algorithms accelerate the simple simultaneous gradient method in the same way.
The theoretical question, where this acceleration does come from, cannot be entirely answered. The use of a direction near $\check{S}$ also in the actual step of iteration, obviously, causes an effect of acceleration. After stating this algorithm in detail, we discuss it in relation to Chebyshev accelerated Simultaneous Iteration.
Let us call this algorithm "W-accelerated simultaneous gradient method".

<u>Algorithm W-Accelerated Simultaneous Gradient Method</u>

(1) $R := \mathcal{A} X - \mathcal{B} X \Delta$.

(2) $A_{11} = \Delta$,

$A_{12} = A_{21}^T := X^H \mathcal{A} R$,

$A_{13} = A_{31}^T := X^H \mathcal{A} W$,

$A_{22} := R^H \mathcal{A} R$,

$A_{23} = A_{32}^T := R^H \mathcal{A} W$,

$A_{33} := W^H \mathcal{A} W$,

$B_{11} = I$,

$B_{12} = B_{21}^T := X^H \mathcal{B} R$,

$B_{13} = B_{31}^T := X^H \mathcal{B} W$,

$B_{22} := R^H \mathcal{B} R$,

$B_{23} = B_{32}^T := R^H \mathcal{B} W$,

$B_{33} := W^H \mathcal{B} W$,

$$A = \begin{pmatrix} A_{11} & A_{12} & A_{13} \\ A_{21} & A_{22} & A_{23} \\ A_{31} & A_{32} & A_{33} \end{pmatrix}, \qquad B \text{ resp.},$$

Solve the (3px3p)-eigenvalue problem of the pencil $\{A,B\}$. Calculate p largest eigenvalues $\hat{\Delta} = \text{diag}(\hat{\delta}_1, \ldots, \hat{\delta}_p)$ with corresponding eigenvectors $Q = (q_1, \ldots, q_p) \in \mathbb{R}^{3p\times p}$,

$$AQ = BQ\hat{\Delta}$$

with

$$Q^T BQ = I.$$

Let Q be divided into three (pxp)-matrices

$$Q = \begin{pmatrix} Q_1 \\ Q_2 \\ Q_3 \end{pmatrix}.$$

(3) $\hat{W} := RQ_2 + WQ_3,$

$$\hat{X} := XQ_1 + \hat{W}.$$

Corollary

1. Obviously, the pencil $\{A,B\}$ represents the Ritz-projection of $\{\mathcal{A},\mathcal{B}\}$ into the 3p-dimensional subspace

$$\text{span}\,(X \vdots R \vdots W),$$

respectively, $\hat{X}$ becomes a Ritz-basis

$$\hat{X} = (X \vdots R \vdots W)Q$$

due to step (3).

2. The algorithm of the conjugate gradient direction required the calculation of the (pxp)-matrices

$$W^H \mathcal{A} R,\ X^H \mathcal{A} R,\ R^H \mathcal{A} R,$$

$$W^H \mathcal{B} R,\ X^H \mathcal{B} R,\ \text{and}\ R^H \mathcal{B} R.$$

Here, the same inner products occur for the definition of A_{12}, A_{23}, A_{22}, B_{12}, B_{23} and B_{22}. The remaining matrices A_{13}, A_{33}, B_{13} and B_{33} are easily obtained by a recurrent calculation in order to avoid an additional increasing of the entire complexity.

Let us denote by

$$Q = \begin{pmatrix} Q_1 \\ Q_2 \\ Q_3 \end{pmatrix} \quad \text{and} \quad \tilde{Q} = \begin{pmatrix} \mathbb{O} \\ Q_2 \\ Q_3 \end{pmatrix}$$

the (3pxp) -matrices defining

$$\hat{W} = (X \vdots R \vdots W)\tilde{Q}$$

and

$$\hat{X} = (X \vdots R \vdots W)Q,$$

then the matrices $\hat{A}_{13}$, $\hat{A}_{33}$, $\hat{B}_{13}$ and $\hat{B}_{33}$ for the next step of the iteration are just determined in the actual step:

$$\hat{B}_{13} = \hat{X}^H \mathcal{B} \hat{W} = Q^T(X:R:W)^H \mathcal{B} (X:R:W)\tilde{Q} = Q^T B\tilde{Q},$$

$$\hat{B}_{33} = \tilde{Q}^T B\tilde{Q},$$

$$\hat{A}_{13} = Q^T A\tilde{Q} = \varDelta \hat{B}_{13},$$

$$\hat{A}_{33} = \tilde{Q}^T A\tilde{Q}.$$

These calculations require $\approx 15p^3$ essential operations only (independent of the large dimension n). Here, the fact

$$\hat{B}_{13} = \hat{B}_{33} + \begin{pmatrix} Q_1 \\ \mathbb{O} \\ \mathbb{O} \end{pmatrix}^T B\tilde{Q}$$

should be used. A convenient start of these recursions consists in the choice $W_0 = \mathbf{0}$ for the first step.
Hence, the computational costs for the calculation of the Ritz-projection agrees for both algorithms. But the step (3), here, needs $3p^2n$ multiplications only, which is to be compared with $4p^2n$ in step (5) and with additional p^2n operations for $S = R + WT$ in the conjugate gradient method.

Remark. The previous algorithm can also be thought to calculate the best p-dimensional subspace span $\hat{X}$ in the 3p-dimensional subspace

$$\text{span } (X \vdots R(X) \vdots \check{X}).$$

Note that $\check{X}$ represents the iterated element of the preceding step.
Our remark follows from step (3):

$$X = \check{X}\check{Q}_1 + W$$

which yields

$$W = X - \check{X}\check{Q}_1.$$

If $\check{Q}_1$ is a regular (pxp)-matrix (obviously true, may be except for some early steps), then we have

$$\begin{aligned} \text{span}\,(X\vdots R\vdots W) &= \text{span}\,(X\vdots R\vdots X-\check{X}\check{Q}_1) \\ &= \text{span}\,(X\vdots R\vdots \check{X}\check{Q}_1) \\ &= \text{span}\,(X\vdots R\vdots \check{X}). \end{aligned}$$

Hence, the basis $(X\vdots R\vdots W)$ used here is a special (numerically better) basis of the subspace spanned by the columns of X, R, and the preceding element $\check{X}$. The presence of $\check{X}$ seems to be one reason for the acceleration of the simple gradient method. For the special case $\beta = \gamma$ a remarkable relation to the Simultaneous Iteration Method with Chebyshev acceleration is obtained. This algorithm will be described in detail in the following Chapter 3.5. Here, we use the resulting iteration formula

$$\hat{X}_{Ch} = (\beta X + \alpha \mathcal{A} X - \check{X}\check{T})T$$

with some fixed constants $\alpha, \beta \in \mathbb{R}$ and two (pxp)-matrices T and $\check{T}$. The constants α and β are defined to form the transformation of an interval $[\underline{\lambda}, \bar{\lambda}]$, containing remaining eigenvalues, into the interval $[-1,1]$. For our purposes, it is sufficient to note that the actual step of the algorithm calculates

$$\hat{X}_{Ch} = \beta XT + \alpha \mathcal{A} XT - XT\check{T}$$

forming one p-dimensional subspace from

$$\text{span}\,(X\vdots \mathcal{A}X\vdots \check{X}).$$

But our W-accelerated gradient method defines that p-dimensional subspace in

$$\text{span}\,(X\vdots R\vdots W) = \text{span}\,(X\vdots R\vdots \check{X}) = \text{span}\,(X\vdots \mathcal{A}X\vdots \check{X}),$$

which guarantees maximal value $\mu(\hat{X})$. Theoretically, this does not mean that our algorithm globally accelerates the method of Simultaneous Chebyshev Iteration, but the numerical examples demonstrate this indeed.

3.4. Algorithms for the Special Eigenproblem of One Operator $\mathcal{A}$, Based on Gradient Directions

There are at least two reasons for the consideration of the special case $\mathcal{B} = \mathcal{I}$, that is, the eigenproblem of the operator $\mathcal{A}$. At first, in the algorithms mentioned above some important simplifications arise from $\mathcal{B} = \mathcal{I}$, which will lead to a further decrease of the total complexity of an iteration step. This advantage, as will be seen later, can also be used in connection with spectral transformations of a given problem.
Secondly, for the special eigenproblem of one operator $\mathcal{A}$ there exists a variety of algorithms, which are based on the power method or on Krylov subspace techniques. Here, we first of all deal with the simultaneous method of steepest ascent and its accelerations according to 3.2. and 3.3. paying attention to $\mathcal{B} = \mathcal{I}$. The other methods, the Simultaneous Iteration (or Subspace Iteration) and the Lanczos algorithms will be discussed in Chapters 3.5. and 3.6., resp.
From Chapter 3.1. we may recall that the simultaneous method of steepest ascent for $\mathcal{B} = \mathcal{I}$ becomes the following algorithm:

(1) $R = \mathcal{A}X - X\Delta.$

(2) $$A = \begin{pmatrix} \Delta & X^H\mathcal{A}R \\ R^H\mathcal{A}X & R^H\mathcal{A}R \end{pmatrix}, \qquad B = \begin{pmatrix} I & X^HR \\ R^HX & R^HR \end{pmatrix}.$$

(3) Solve $AQ = BQ\hat{\Delta}$ (p largest eigenvalues).

(4) $\hat{X} = (X \vdots R)Q.$

Here, the fact $\mathcal{B} = \mathcal{I}$ leads to some simplifications. Due to (2.4.12) we have

$$B_{12} = B_{21}^T = X^HR = \mathbb{O}.$$

On the other hand, the results $\hat{\Delta}$ and $\hat{X}$ of that step of the iteration depend on the subspace span $(X \vdots R)$ only. For that subspace a more simple basis without calculating R is known

$$\text{span}\,(X \vdots R) = \text{span}\,(X \vdots \mathcal{A}X).$$

From the numerical point of view the basis $(X \vdots \mathcal{A}X)$ seems to be more dangerous than $(X \vdots R)$. If X converges to Φ_1, some columns of $(X \vdots \mathcal{A}X)$ become more and more parallel. Here, during the solution process of the small eigenvalue problem $\{A,B\}$ numerical instabilities can be avoided by resorting to a relatively easy procedure. The relevant questions pertaining to the implementation will be described in Chapter 6. If the basis $(X \vdots \mathcal{A}X)$ for the actual subspace is used, the algorithm is changed into the following

<u>Algorithm of Steepest Ascent for Largest Eigenvalues of $\mathcal{A}$</u>

(1) $Z := \mathcal{A}X$.

(2) Calculate $E_2 = X^H \mathcal{A}^2 X = Z^H Z$

and $E_3 = X^H \mathcal{A}^3 X = Z^H \mathcal{A} Z$

then
$$A = \begin{pmatrix} \Delta & E_2 \\ E_2 & E_3 \end{pmatrix}$$
$$B = \begin{pmatrix} I & \Delta \\ \Delta & E_2 \end{pmatrix},$$

solve $AQ = BQ\hat{\Lambda}$ (p largest eigenvalues).

(3) $\hat{X} = (X \vdots Z)Q$.

It should be noted that the calculation of E_3 does not imply the multiplication $\mathcal{A}Z$, if a proper inner product is defined. The special inner product depends on the definition of the symmetric operator $\mathcal{A}$ and has to be discussed in connection with the application to a matrix eigenvalue problem only, which will be done in Chapter 5.
This algorithm represents an immediate acceleration of the well-known Simultaneous Iteration Method. The details of those algorithms will be found in the next chapter. We emphasize here the fact that $\hat{X}_{SI}$ is calculated to be a special basis of span $(\mathcal{A}X)$ in the actual step of that iteration process. Here, span $\hat{X}$ has the largest functional value $\mu(\hat{X})$ over span $(X \vdots \mathcal{A}X)$, thus

$$\mu(\hat{X}_{SI}) \leq \mu(\hat{X}).$$

In Chapter 4 we prove that the rate of convergence of the gradient method coincides with the rate of convergence of a Simultaneous Iteration with best shifted operator

$$\mathcal{A} - \frac{\lambda_{p+1} + \lambda_n}{2} \mathcal{I}.$$

Hence, also globally, a better behaviour of the above algorithm is evident. In the more sophisticated accelerations of this method of steepest ascent obtained in Chapter 3.2. and 3.3. the total complexity is also decreased for the special case $\mathcal{B} = \mathcal{I}$.
For the simultaneous conjugate gradient method

$$\text{span } (X \vdots R + WT) = \text{span } (X \vdots \mathcal{A}X + WT)$$

and for the W-accelerated gradient method

$$\text{span } (X \vdots R \vdots W) = \text{span } (X \vdots \mathcal{A}X \vdots W)$$

can be concluded from $\mathcal{B} = \mathcal{I}$. Hence, for both algorithms the calculation of

$$E_2 = X^H \mathcal{A}^2 X \quad \text{and}$$

$$E_3 = X^H \mathcal{A}^3 X$$

together with only one further (pxp)-matrix

$$E_W = W^H \mathcal{A}^2 X$$

is sufficient for the specification of the Ritz-projection $\{A,B\}$, as will be seen.

Simultaneous Conjugate Gradient Method for the Largest Eigenvalues of $\mathcal{A}$

(1) $Z := \mathcal{A}X.$

(2) $E_2 := Z^H Z = X^H \mathcal{A}^2 X,$

$E_3 := Z^H \mathcal{A} Z = X^H \mathcal{A}^3 X,$

$E_W := W^H \mathcal{A}^2 X,$

calculate T from

$$T \Delta_1 - \Delta_2 T = E_W$$

using the remarks (3.2.7)/(3.2.8).

(3) $S = Z + WT.$

(4) $A_{11} = \Delta_1,$

$A_{12} = A_{21}^T = E_2,$

$A_{22} = E_3 + T^T E_W + E_W^T T + T^T \Delta_2 T,$

$B_{11} = I,$

$B_{12} = B_{21}^T = \Delta_1,$

$B_{22} = E_2 + T^T T,$

$$A = \begin{pmatrix} A_{11} & A_{12} \\ A_{21} & A_{22} \end{pmatrix}, \quad B \text{ resp.}$$

Solve the full eigenproblem for $\{A, B\}$

$AQ = BQ\hat{\Delta},$

$\hat{\Delta} = \mathrm{diag}\,(\hat{\Delta}_1, \hat{\Delta}_2)$ ($\hat{\Delta}_1$ contains p largest, $\hat{\Delta}_2$ p smallest eigenvalues).

(5) $(\hat{X} \vdots \hat{W}) = (X \vdots S)Q.$

W-Accelerated Simultaneous Gradient Method for the Largest Eigenvalues of $\mathcal{A}$

(1) $Z := \mathcal{A} X$.

(2) $E_2 := Z^H Z = X^H \mathcal{A}^2 X$,

$E_3 := Z^H \mathcal{A} Z = X^H \mathcal{A}^3 X$,

$E_W := W^H \mathcal{A}^2 X$.

(3) $A_{11} = \Delta$,

$A_{12} = A_{21} := E_2$,

$A_{22} := E_3$,

$A_{23} = A_{32}^T := E_W^T$,

$B_{11} = I$,

$B_{12} = B_{21} = \Delta$,

$B_{22} := E_2$.

(The matrices $A_{13} = A_{31}^T = B_{23} = B_{32}^T$, $B_{13} = B_{31}^T$, A_{33}, and B_{33} are given from the preceding step.)

$$A = \begin{pmatrix} A_{11} & A_{12} & A_{13} \\ A_{21} & A_{22} & A_{23} \\ A_{31} & A_{32} & A_{33} \end{pmatrix}, \quad B \text{ resp.}$$

Solve $AQ = BQ\hat{\Delta}$ ($\hat{\Delta}$ contains p largest eigenvalues),

$Q^T BQ = I$,

$$Q = \begin{pmatrix} Q_1 \\ Q_2 \\ Q_3 \end{pmatrix}, \qquad \tilde{Q} = \begin{pmatrix} 0 \\ Q_2 \\ Q_3 \end{pmatrix}.$$

(4) $\hat{W} := ZQ_2 + WQ_3$,

$\hat{X} := XQ_1 + \hat{W}$,

$\hat{A}_{13} := Q^T A\tilde{Q}$, $\hat{A}_{33} := \tilde{Q}^T A\tilde{Q}$,

$\hat{B}_{13} := Q^T B\tilde{Q}$, $\hat{B}_{33} := \tilde{Q}^T B\tilde{Q}$.

3.5. The Simultaneous Iteration Method

This method was developed by [BAUER 1] in 1967 and [RUTISHAUSER 1, 2]. Since that time a huge body of literature, where the details, the convergence properties and the application are discussed, has been published. The most important work on this topic is due to BATHE/WILSON, JENNINGS, STEWART and McCORMICK/NOE .
For a number of reasons this method was widely used in many branches of engineering, especially for the investigation of eigenoscillation problems. One reason consists in the fact that this algorithm needs two basic operations only. Another important reason is the famous numerical behaviour. This method will be found to be considerably stable. A proper implementation provided, the method does not involve any major difficulties.
At the beginning of the eighties, however, some criticism of that method has emerged after the latest development of Lanczos algorithms. Nevertheless, the Simultaneous Iteration Method is also applicable to nonsymmetric operators - here, Lanczos-like algorithms (or, more precisely: Krylov subspace methods) are still in a stage of development.
All variants of the Simultaneous Iteration can be described as an algorithm consisting of three steps:

Simultaneous Iteration Method for absolutely largest eigenvalues of $\mathcal{A}$

(1) $Z := \mathcal{A}X$.

(2) Calculate T.

(3) $\hat{X} = ZT$.

Corollary. Provided that each T is nonsingular, then the k-th iterated element X_k fulfils the following formula

$$X_k = \mathcal{A}^k X_o T_o \ \cdots \ T_{k-1}, \tag{3.5.1}$$

thus,

$$\text{span } X_k = \text{span } \mathcal{A}^k X_o. \tag{3.5.2}$$

The fact (3.5.2) alone will lead to the proof of the convergence of this method in the sense

$$\text{span } X_k \longrightarrow \text{span } \Phi_1 \ .$$

Here, the functional $\mu(X)$ is negligible for the proofs of convergence, nevertheless an increase of $\mu(X_k)$ can also be proved under some assumptions.
One main difference between this method and the previously described algorithms consists in the fact that the largest eigenvalues in absolute magnitude (together with corresponding eigenvectors) will be calculated

here. For purposes of easier comparison of the methods we assume for the beginning that $\mathcal{A}$ is a symmetric positive definite operator, hence the largest eigenvalues $\lambda_1 \geq \dots \geq \lambda_p$ are also the absolutely largest. Then the following lemma holds:

Lemma 3.5.1. If $\mathcal{A}$ is positive definite symmetric, then the elements X_k obtained in the Simultaneous Iteration Method, satisfy

$$\mu(X_{k+1}) \geq \mu(X_k).$$

Proof. We drop the index k of iteration again, that is, for

$$\hat{X} = \mathcal{A} X T$$

we prove

$$\mu(\hat{X}) \geq \mu(X).$$

Obviously, T is assumed to be regular, thus

$$\mu(\mathcal{A} X) = \mu(\hat{X}).$$

From the definition

$$\mu(X) = \operatorname{tr}\,(X^H X)^{-1}(X^H \mathcal{A} X) = \sum_{i=1}^{p} \delta_i$$

holds, where δ_i denote the eigenvalues of the pencil $\{X^H \mathcal{A} X, X^H X\}$ (the Ritz-eigenvalues of span X).
Hence,

$$\mu(\hat{X}) = \mu(\mathcal{A} X) = \sum_{i=1}^{p} \hat{\delta}_i$$

and $\hat{\delta}_i$ are the eigenvalues of the pencil $\{X^H \mathcal{A}^3 X,\ X^H \mathcal{A}^2 X\}$. The inequality

$$\frac{t^T\ X^H \mathcal{A}^3 X\ t}{t^T\ X^H \mathcal{A}^2 X\ \ t} \geq \frac{t^T\ X^H \mathcal{A} X\ t}{t^T\ X^H X\ t} \qquad \forall\ t \in \mathbb{R}^p \tag{3.5.3}$$

would imply $\hat{\delta}_i \geq \delta_i$ $(i=1,\dots,p)$ and would complete the proof. This inequality is an immediate consequence of Cauchy's inequality,

$$\langle \mathcal{A} x, x\rangle^2 \leq \langle \mathcal{A}^2 x, x\rangle\ \langle x, x\rangle,$$

which (for $\mathcal{A}^{\frac{1}{2}} x$ instead of x) leads to

$$\langle \mathcal{A}^2 x, x\rangle^2 \leq \langle \mathcal{A}^3 x, x\rangle \langle \mathcal{A} x, x\rangle .$$

Thus,

$$\frac{\langle \mathcal{A}^3 x, x\rangle}{\langle \mathcal{A}^2 x, x\rangle} \geq \frac{\langle \mathcal{A}^2 x, x\rangle}{\langle \mathcal{A} x, x\rangle} \geq \frac{\langle \mathcal{A} x, x\rangle}{\langle x, x\rangle} \quad .$$

For x = Xt the inequality (3.5.3) is obtained.∎

For a more thorough study of this method we emphasize here that the operator $\mathcal{A}$ need not be positive definite; it need not be symmetric either. Thus (for the next considerations only) the operator $\mathcal{A}$ will not be assumed to be symmetric and has eigenvalues $\lambda_1, \ldots, \lambda_n$, which could be complex numbers. Here, they will be labelled as

$$|\lambda_1| \geq \ldots \geq |\lambda_p| \geq |\lambda_{p+1}| \geq \ldots \geq |\lambda_n| .$$

For purposes of brevity we assume $\mathcal{A}$ to be non-defective, that is, there exists a full basis of eigenvectors Φ,

$$\mathcal{A}\, \Phi = \Phi \Lambda$$

(The generalization to defective operators $\mathcal{A}$ is straightforward, then Λ is replaced by a Jordan canonical form, for details see [MEYER 7].) Together with Φ, for a non-symmetric (especially non-normal) operator $\mathcal{A}$ also left eigenelements

$$\Psi = (\Psi_1 \vdots \Psi_2) = (\psi_1, \psi_2, \ldots, \psi_n)$$

are of interest,

$$\Psi_1 \in \mathbf{H}^p, \ \Psi_2 \in \mathbf{H}^{n-p},$$

$$\mathcal{A}^* \Psi = \Psi \bar{\Lambda} ,$$

$$\Psi^H \Phi = I$$

Because of the fact that the convergence properties of this method are well-known, we add the convergence theorem here in this chapter. Chapter 4 will be concerned with the new algorithms of 3.1. and 3.2. and 3.3. only, for which the investigations of the convergence does not coincide with those for Simultaneous Iteration.

Theorem 3.5.1. If $|\lambda_p| > |\lambda_{p+1}|$ (3.5.4)

and if the matrix

$$H_o = \Psi_1^H X_o \ \text{is regular,} \qquad (3.5.5)$$

then span X_k converges to span Φ_1 (provided T_k is regular for any step of the iteration).

Proof. In the eigenexpansion of X_o,

$$X_o = \Phi V_o,$$
$$V_o = \Psi^H X_o \in \mathbb{C}^{nxp},$$

the matrix V_o can be divided into two parts

$$V_o = \begin{pmatrix} H_o \\ G_o \end{pmatrix}, \quad H_o \in \mathbb{C}^{pxp}, \; G_o \in \mathbb{C}^{(n-p)xp}$$

With (3.5.1) this leads to

$$X_k = \Phi \Lambda^k V_o T_o \, \dots \, T_{k-1}$$

and

$$\begin{aligned} \text{span } X_k &= \text{span } (\Phi \Lambda^k V_o) \\ &= \text{span } (\Phi_1 \Lambda_1^k H_o + \Phi_2 \Lambda_2^k G_o) \\ &= \text{span } (\Phi_1 + \Phi_2 \Lambda_2^k G_o H_o^{-1} \Lambda_1^{-k}) \end{aligned}$$

From the assumption (3.5.4) we conclude the convergence of the matrix $\Lambda_2^k G_o H_o^{-1} \Lambda_1^{-k}$ to zero, thus

$$\text{span } X_k \longrightarrow \text{span } \Phi_1,$$

and the rate of this convergence is linear with the quotient $\eta = |\lambda_{p+1}| / |\lambda_p|$:

$$\| \Lambda_2^k G_o H_o^{-1} \Lambda_1^{-k} \|_2 \leq \| G_o H_o^{-1} \|_2 \cdot \eta^k.$$

■

Up to this point, the special choice of the (pxp)-matrix T in the iteration steps was not discussed. According to Theorem 3.5.1 this matrix cannot affect the convergence of span X_k, but it will certainly affect the special basis $\hat{X}$ of the next subspace. Furthermore, together with the calculation of T the approximations of the eigenvalues belonging to that step are obtained (for most cases). Obviously, the matrix T has to ensure that the basis X_k remains numerically "good enough", that is, it has to avoid the columns of X_k becoming more and more parallel. The single variants of the Simultaneous Iteration Method are thus distinguished by the special choice of T. These variants are considered in [PARLETT 1] and [MEYER 3].

Here, we recommend one variant with considerably sparse computational costs together with the appropriate convergence of eigenvalue and eigenvector approximations.

For a symmetric operator $\mathcal{A}$ the special basis $\hat{X}$ could be chosen to be the Ritz-basis of

$$\text{span } (\mathcal{A} X) = \text{span } Z.$$

Then, T would contain the eigenvectors of the Ritz-projection $\{Z^H\mathcal{A}Z, Z^HZ\}$. If the operator $\mathcal{A}$ is non-symmetric, but <u>normal</u> ($\mathcal{A}\mathcal{A}^* = \mathcal{A}^*\mathcal{A}$) this pencil could lead to complex arithmetic, which should be avoided as best as possible. Due to the fact that $\mathcal{A}$, $\mathcal{A}^*$ and $\mathcal{A}^*\mathcal{A}$ (and so on) have the same eigenelements, there is no need to project the operator $\mathcal{A}$ itself into span Z. We can also define $\hat{X}$ to be the Ritz-basis of span Z with respect to the operator $f(\mathcal{A})$ (a function of $\mathcal{A}$). Furthermore, if $\mathcal{A}$ is a normal operator (with respect to the given inner product in $\mathbf{H}$), then $\mathcal{A}$ is also normal with $(\mathcal{A}\mathcal{A}^*)$- or with $(\mathcal{A}\mathcal{A}^*)^{-1}$- inner product. The Ritz-step for $f(\mathcal{A}) = \mathcal{A}\mathcal{A}^*$ using $(\mathcal{A}\mathcal{A}^*)^{-1}$-inner product leads to the pencil

$$\{Z^H(\mathcal{A}\mathcal{A}^*)^{-1}(\mathcal{A}\mathcal{A}^*)Z, Z^H(\mathcal{A}\mathcal{A}^*)^{-1}Z\} = \{Z^HZ,\ X^HX\}.$$

If $X^HX = I$ could be guaranteed, the definition of the Ritz-step requires the calculation of one symmetric (pxp)-matrix $A = Z^HZ$, only. The resulting algorithm is the following. The Ritz-eigenvalues will be denoted by δ_i^2, because these numbers converge to $|\lambda_i|^2$ for normal operators.

<u>Algorithm of Simultaneous Iteration for absolutely largest eigenvalues of an operator $\mathcal{A}$</u>

(1) $Z := \mathcal{A}X$.

(2) $A := Z^HZ$.

Solve the full eigenproblem of the matrix A:

$AQ = Q\hat{\Delta}^2$, $Q^TQ = I$ (eigenvalues: $\hat{\delta}_1^2 \geq \ldots \geq \hat{\delta}_p^2$).

(3) $\hat{X} = ZQ\hat{\Delta}^{-1}$.

Note that the definition $T = Q\hat{\Delta}^{-1}$ indeed yields

$$\hat{X}^H\hat{X} = \hat{\Delta}^{-1}Q^TZ^HZQ\hat{\Delta}^{-1} = I,$$

thus $\hat{\Delta}^2$ contains Ritz-eigenvalues of the subspace span $\hat{X}$ for the operator $\mathcal{A}\mathcal{A}^*$ with respect to the $(\mathcal{A}\mathcal{A}^*)^{-1}$-inner product. This fact remains also true for non-normal operators. [McCORMICK/NOE 1] investigated this algorithm for normal operators and proved the following convergence theorem for the eigenvalue approximations.

<u>Theorem 3.5.2.</u> For normal operator $\mathcal{A}$ the eigenvalue approximations $\hat{\delta}_i^2$ at the k-th step of iteration can be estimated as

$$\hat{\delta}_i^2 \leq |\lambda_i|^2 \leq \hat{\delta}_i^2(1 + \varepsilon_{ik}), \qquad i=1,\ldots,p,$$

where

$$\varepsilon_{ik} \leq \left|\frac{\lambda_{p+1}}{\lambda_i}\right|^{2k} \cdot \|G_oH_o^{-1}\|_2 .$$

Here, we emphasize that this algorithm also should be used for the case

of non-normal operators $\mathcal{A}$. For the first view this seems unusual, and some remarks must be added here.
For a normal operator the Ritz-technique described above will guarantee that the columns of X converge to eigenvectors (for simple eigenvalues λ_i, more precisely, for simple complex pairs of eigenvalues). For non-normal operators this convergence is impossible, because X represents an orthonormal basis at any step, whereas the eigenvectors Φ_1 do not. Hence, for non-normal $\mathcal{A}$ the algorithm produces a series of orthonormal bases X of the subspaces converging to the dominant "eigen subspace" span Φ_1. For the construction of these bases at any step a symmetric eigenvalue problem of the matrix A occurs. The eigenvalues δ_i^2 of A will not converge to eigenvalues of $\mathcal{A}^*\mathcal{A}$ but these numbers can be used to indicate the step, when the subspace span X approximates span Φ_1 with expected accuracy. At this point the iteration process is stopped and the eigenvectors Φ_1 can be calculated from span X with the help of a non-symmetric matrix problem, as will be seen later. Here, we are going to prove that the numbers δ_i are convergent, thus they can be used for a stopping criterion.

<u>Lemma 3.5.2.</u> For arbitrary operators $\mathcal{A}$ the eigenvalues δ_i^2 of the matrix A in the k-th step of the algorithm described above converge to p numbers $\epsilon_1^2 \geq \ldots \geq \epsilon_p^2$, which are the eigenvalues of the matrix pencil

$$\{\bar{\Lambda}_1 \Phi_1^H \Phi_1 \Lambda_1 , \quad \Phi_1^H \Phi_1 \}$$

<u>Proof.</u> From the algorithm defined above the following formulas are deduced

$$X_k = \mathcal{A}^k X_o T_o \ldots T_{k-1},$$

$$Z_k = \mathcal{A} X_k = \mathcal{A}^{k+1} X_o T_o \ldots T_{k-1}.$$

The eigenvalues of the matrix $A_k = Z_k^H Z_k$ respectively, the eigenvalues of the matrix pencil $\{A_k, I\} = \{Z_k^H Z_k, X_k^H X_k\}$ are considered. Using the eigenexpansion of X_o, we transform the pencil $\{A_k, I\}$ into the equivalent pencil $\{\tilde{A}_k, \tilde{B}_k\}$, which obviously converges to $\{\bar{\Lambda}_1 \Phi_1^H \Phi_1 \Lambda_1, \Phi_1^H \Phi_1\}$, as will be seen.
With

$$X_o = \Phi V_o$$

we have

$$X_k^H X_k = T_{k-1}^T \ldots T_o^T V_o^* \bar{\Lambda}^k \Phi^H \Phi \Lambda^k V_o T_o \ldots T_{k-1}$$

and

$$Z_k^H Z_k = T_{k-1}^T \dots T_o^T V_o^* \bar{\Lambda}^{k+1} \Phi^H \Phi \Lambda^{k+1} V_o T_o \dots T_{k-1}.$$

According to the nonsingularity of T_j the pencil $\{Z_k^H Z_k, X_k^H X_k\}$ is equivalent to

$$\{V_o^* \bar{\Lambda}^{k+1} \Phi^H \Phi \Lambda^{k+1} V_o,\; V_o^* \bar{\Lambda}^k \Phi^H \Phi \Lambda^k V_o\}.$$

If V_o is again divided into

$$V_o = \begin{pmatrix} H_o \\ G_o \end{pmatrix}, \qquad H_o \in \mathbb{C}^{p\times p},$$

then

$$\Phi \Lambda^k V_o = \Phi_1 \Lambda_1^k H_o + \Phi_2 \Lambda_2^k G_o .$$

Thus, multiplying the last pencil by $(\Lambda_1^k H_o)^{-1}$ from the right (its adjoint matrix from the left), it is transformed into $\{\tilde{A}_k, \tilde{B}_k\}$ with

$$\tilde{A}_k = \bar{\Lambda}_1 \Phi_1^H \Phi_1 \Lambda_1 + U_k^* \bar{\Lambda}_2 \Phi_2^H \Phi_1 \Lambda_1 + \bar{\Lambda}_1 \Phi_1^H \Phi_2 \Lambda_2 U_k + U_k^* \bar{\Lambda}_2 \Phi_2^H \Phi_2 \Lambda_2 U_k,$$

$$\tilde{B}_k = \Phi_1^H \Phi_1 + U_k^* \Phi_2^H \Phi_1 + \Phi_1^H \Phi_2 U_k + U_k^* \Phi_2^H \Phi_2 U_k$$

and

$$U_k = \Lambda_2^k G_o H_o^{-1} \Lambda_1^{-k} \rightarrow \mathbb{O}$$

Hence, the eigenvalues of A_k are the eigenvalues of $\{\tilde{A}_k, \tilde{B}_k\}$ too and converge to $\sigma_1^2 \dots \sigma_p^2$. The rate of this convergence depends on $\eta = |\lambda_{p+1}/\lambda_p|$ and is thus identical to that of Theorem 3.5.1 for span $X_k \rightarrow$ span Φ_1. ∎

<u>Corollary.</u> The algorithm of Simultaneous Iteration in the above variant was developed for normal operators, but can also be used for arbitrary $\mathcal{A}$. Here, at least two advantages should be noted. At first, for a real operator $\mathcal{A}$ the complex arithmetic is entirely avoided during the iteration process. Secondly, the computation of the orthonormal basis $\hat{X}$ requires the solution of a symmetric (pxp)-eigenvalue problem only (which is numerically much cheaper and more stable than unsymmetric problems).
If the iteration is stopped due to the information that the eigenvalues of A do not change (up to the expected accuracy), then the last basis X represents an orthonormal basis of the subspace

$$\text{span } X = \text{span } \Phi_1$$

(approximately). Hence, there exists an (pxp)-matrix U with

$$\Phi_1 = XU. \tag{3.5.6}$$

The computation of U requires the solution of a non-symmetric eigenvalue problem: From (3.5.6) we have

$$\mathcal{A} XU = \mathcal{A} \Phi_1 = \Phi_1 \Lambda_1 = XU \Lambda_1$$

thus

$$X^H \mathcal{A} XU = U \Lambda_1$$

that is, U is obtained as eigenvector matrix of the non-symmetric matrix $X^H \mathcal{A} X$, which is to be calculated after the process of iteration.
Hence, the solution of a non-symmetric (pxp)-eigenvalue problem occurs only once, i.e. after the end of the iteration.
For a normal operator $\mathcal{A}$ at least the calculation of the matrix $X^H \mathcal{A} X$ after the iteration process is important to get eigenvalues λ_i (only $\sigma_i = |\lambda_i|$ are the results of the iteration).
Even though this method has been widely discussed here, it could only be recommended for non-symmetric operators.
For symmetric operators some other algorithms has been developed over the last ten years, which will essentially decrease the necessary computational costs. Especially the algorithms of Chapter 3.4. and Lanczos-algorithms will be superior to the simple Simultaneous Iteration Method and even superior to Simultaneous Iteration with Chebyshev acceleration. This method was developed by [RUTISHAUSER 1] and will be described(in a slightly improved manner) in the following.
We will restrict ourselves to a symmetric operator $\mathcal{A}$ again. If an interval $[\underline{\lambda}, \bar{\lambda}]$ is known, which contains all unwanted eigenvalues of $\mathcal{A}$

$$\lambda_i \in [\underline{\lambda}, \bar{\lambda}], \qquad i = p+1, \ldots, n,$$

and does not contain the eigenvalues $\lambda_1, \ldots, \lambda_p$, then an acceleration of Simultaneous Iteration Method with the help of Chebyshev polynomials can be started. Note that $\lambda_1, \ldots, \lambda_p$ need not be the absolutely largest eigenvalues (as in the case of the usual Simultaneous Iteration Method).
If there exists a polynomial $f_k(\tau)$ in τ of degree k with

$$|f_k(\lambda_i)| < 1 \text{ for } i = p+1, \ldots, n \tag{3.5.7}$$

and

$$|f_k(\lambda_i)| > 1 \text{ for } i = 1, \ldots, p, \tag{3.5.8}$$

then the convergence span $X_k \to \operatorname{span} \Phi_1$ obviously remains valid, if formula (3.5.1) is replaced by

$$X_k = f_k(\Lambda) X_o T_o \ldots T_{k-1}. \tag{3.5.9}$$

In the proof of Theorem 3.5.1 the matrix Λ^k has to be replaced by

$$f_k(\Lambda) = \operatorname{diag}\,(f_k(\lambda_1), \ldots, f_k(\lambda_n))$$

and the convergence quotient

$$\eta^k = \left| \frac{\lambda_{p+1}}{\lambda_p} \right|^k$$

changes to

$$\frac{\max\limits_{i>p} |f_k(\lambda_i)|}{\min\limits_{i\leq p} |f_k(\lambda_i)|} \leq \frac{1}{\min\limits_{i\leq p} |f_k(\lambda_i)|} < 1$$

(which should converge to zero).
With the help of Chebyshev polynomials and their recurrence relations the formula (3.5.9) is transformed into the algorithm of Simultaneous Chebyshev Iteration.
Therefore, we define

$$g(\tau) = \alpha'\tau + \beta'$$

which transforms the interval $[\underline{\lambda}, \overline{\lambda}]$ into $[-1, 1]$, that is

$$\alpha' = \frac{2}{\overline{\lambda} - \underline{\lambda}} = \frac{\alpha}{2}, \qquad \beta' = -\frac{\overline{\lambda} + \underline{\lambda}}{\overline{\lambda} - \underline{\lambda}} = \frac{\beta}{2}. \tag{3.5.10}$$

Let $t_k(\tau)$ be the k-th Chebyshev polynomial of first kind in τ,

$$t_o(\tau) = 1, \qquad t_1(\tau) = \tau,$$

$$t_{k+1}(\tau) = 2\tau\, t_k(\tau) - t_{k-1}(\tau) \tag{3.5.11}$$

then

$$f_k(\tau) = t_k(g(\tau))$$

fulfils the assumptions (3.5.7)/(3.5.8) and $|f_k(\lambda_i)|^{-1}$ (for $i \leq p$) tends very rapidly to zero.
From (3.5.11)

$$f_{k+1}(\tau) = (2\alpha'\tau + 2\beta')f_k(\tau) - f_{k-1}(\tau) = (\alpha\tau+\beta)f_k(\tau) - f_{k-1}(\tau)$$

the algorithm is deduced, using the facts

$$\begin{aligned} X_{k+1} &= f_{k+1}(\mathcal{A})\, X_o T_o \dots T_k \\ &= (\alpha\mathcal{A}+\beta)\, f_k(\mathcal{A})\, X_o T_o \dots T_k - f_{k-1}(\mathcal{A})\, X_o T_o \dots T_k \\ &= (\alpha\mathcal{A}+\beta\mathfrak{I})\, X_k T_k - X_{k-1}T_{k-1}T_k . \end{aligned}$$

From the comparison with the simple Simultaneous Iteration we have

$$X_{k+1} = Z_k T_k,$$

thus the algorithm can be completed to

<u>Algorithm of Simultaneous Iteration with Chebyshev Acceleration</u>

(1) $Z := \alpha\mathcal{A}X + \beta X - \check{X}\check{T}$.

(2) calculate T:

$A := Z^H Z$,

solve $AQ = Q\hat{\Delta}^2$ with $Q^TQ = I$,

$T := Q\hat{\Delta}^{-1}$.

(3) $\hat{X} = ZT$.

<u>Remark.</u> If an appropriate interval $[\underline{\lambda},\overline{\lambda}]$ has been obtained and the Chebyshev acceleration, described here, has been started, there is no reason to stop this process. Especially, X always represents an orthonormal basis (more precisely, X is constructed as Ritz-basis of span Z belonging to the operator $\mathcal{A}\mathcal{A}^* = \mathcal{A}^2$ with respect to $\mathcal{A}^{-2}$-inner product, as discussed above). This is the main difference between our algorithm and [RUTISHAUSER's 1] (the implementation in the well-known program RITZIT). In RITZIT the algorithm consists in the formula

$$\hat{X} = \alpha\mathcal{A}X + \beta X - \check{X}$$

only, thus after some steps the Chebyshev iteration has to be stopped for orthonormalizing the columns of X. Our algorithm leads to Chebyshev polynomials of arbitrary high degrees.

Again the eigenvalues of A are convergent numbers and will be used for a stopping criterion:

Lemma 3.5.2. The numbers δ_i^2 calculated at the k-th step of the Simultaneous Chebyshev Iteration (the eigenvalues of A_k) converge to

$$(|\nu_i| + \sqrt{\nu_i^2 - 1})^2, \quad i = 1, \ldots, p,$$

where $\nu_i = g(\lambda_i) = \alpha' \lambda_i + \beta'$.

Proof. Again, some eigenvalue problems equivalent to $\{A_k, I\} = \{Z_k^H Z_k, X_k^H X_k\}$ have to be considered. From (3.5.9) we recall

$$X_k = f_k(\Lambda) X_o T_o \ldots T_{k-1} = \Phi f_k(\Lambda) V_o T_o \ldots T_{k-1},$$

$$Z_k = f_{k+1}(\Lambda) X_o T_o \ldots T_{k-1} = \Phi f_{k+1}(\Lambda) V_o T_o \ldots T_{k-1}$$

(again $X_o = \Phi V_o = \Phi_1 H_o + \Phi_2 G_o$).
Hence, $\{A_k, I\}$ is equivalent to

$$\{ V_o^T f_{k+1}^2(\Lambda) V_o, \; V_o^T f_k^2(\Lambda) V_o \}$$

and by multiplying it with $(f_k(\Lambda_1) H_o)^{-1}$ (from the right, its transpose from the left) this is also equivalent to $\{\tilde{A}_k, \tilde{B}_k\}$, where

$$\tilde{A}_k = f_{k+1}^2(\Lambda_1) f_k^{-2}(\Lambda_1) + \tilde{U}_k^T \tilde{U}_k,$$

$$\tilde{B}_k = I + U_k^T U_k,$$

$$\tilde{U}_k = f_{k+1}(\Lambda_2) G_o H_o^{-1} f_k^{-1}(\Lambda_1),$$

$$U_k = f_k(\Lambda_2) G_o H_o^{-1} f_k^{-1}(\Lambda_1).$$

Obviously, $U_k \to 0$ for $k \to \infty$ due to the formulas (3.5.7)/(3.5.8). Analogous, the matrix

$$\tilde{U}_k = U_{k+1} f_k(\Lambda_1)^{-1} f_{k+1}(\Lambda_1)$$

converges to zero, because we are going to calculate the limit of the diagonal matrix

$$f_{k+1}(\Lambda_1) f_k^{-1}(\Lambda_1).$$

Hence, the eigenvalues of A_k are the eigenvalues of $\{\tilde{A}_k, \tilde{B}_k\}$ and converge to the squares of that limit. Let λ be one of the eigenvalues $\lambda_1, \ldots, \lambda_p$

(contained in Λ_1), thus

$$\lambda \notin [\underline{\lambda}, \bar{\lambda}]$$

respectively

$$\nu = g(\lambda) \text{ with } |\nu| > 1,$$

then we have to determine the limit of

$$f_{k+1}(\lambda)/f_k(\lambda) = t_{k+1}(\nu)/t_k(\nu),$$

which is easily deduced from the formula

$$t_k(\nu) = \frac{1}{2}((\nu + \sqrt{\nu^2-1})^k + (\nu + \sqrt{\nu^2-1})^k) \tag{3.5.12}$$

to be

$$\lim_{k \to \infty} t_{k+1}(\nu)/t_k(\nu) = \begin{cases} \nu + \sqrt{\nu^2-1}, & \nu > 1, \\ \nu - \sqrt{\nu^2-1}, & \nu < -1, \end{cases}$$

$$= \operatorname{sign} \nu \, (|\nu| + \sqrt{\nu^2-1}).$$

■

At the end of this chapter, some remarks on the application of the Chebyshev acceleration for non-symmetric operators should be added. The inclusion of the eigenvalues $\lambda_{p+1}, \ldots \lambda_n$ in the interval $[\underline{\lambda}, \bar{\lambda}]$ can be replaced by an elliptical inclusion region for unwanted eigenvalues. The focuses of such an ellipse, the numbers $\underline{\lambda}$ and $\bar{\lambda}$, lead again to the definition of a transformed Chebyshev polynomial. If $\underline{\lambda}$ and $\bar{\lambda}$ are real numbers, the iteration process remains real again. The convergence to $\lambda_1, \ldots, \lambda_p$ is then proved similar to previous theorems using the following fact:

Let $\mathbb{E}(\beta)$ be the boundary of an ellipse of the complex plane with focuses in -1 and 1 and the semiminor axis β. Let $z_{1/2}$ two complex numbers with

$$z_1 \in \mathbb{E}(\beta_1), \quad z_2 \in \mathbb{E}(\beta_2)$$

and

$$\beta_1 > \beta_2$$

then

$$\lim_{k\to\infty} \frac{t_k(z_2)}{t_k(z_1)} = 0.$$

This is proved by an immediate calculation of $t_k(z)$ with $z = \sqrt{1+\beta^2} \cdot \cos\varphi + i\beta\sin\varphi$ (i.e. $z \in \mathbf{E}(\beta)$) using formula (3.5.12).
Hence, if the ellipse $\mathbf{E}(\underline{\lambda},\overline{\lambda},\beta)$ with focuses $\underline{\lambda}$ and $\overline{\lambda}$ and with semiminor axis

$$\beta \cdot \frac{\overline{\lambda}-\underline{\lambda}}{2}, \qquad \underline{\lambda},\overline{\lambda} \in \mathbf{R},$$

contains the eigenvalues $\lambda_{p+1},\ldots,\lambda_n$ and does not contain $\lambda_1,\ldots,\lambda_p$, then the same transformation (3.5.10)

$$g(z) = \alpha' z + \beta'$$

can be used. Here, $g(z)$ transforms the ellipse $\mathbf{E}(\underline{\lambda},\overline{\lambda},\beta)$ into the unit-ellipse $\mathbf{E}(\beta)$, and the polynomial

$$f_k(\tau) = t_k(g(\tau))$$

leads to the same algorithm as in the symmetric case. Real arithmetic is kept (for real operator $\mathcal{A}$), if $\underline{\lambda}$ and $\overline{\lambda}$ are found as real numbers.
If the Chebyshev polynomial $t_k(\tau)$ is changed into a similar polynomial, defined according to

$$\tilde{t}_0(\tau) = 1, \qquad \tilde{t}_1(\tau) = \tau,$$

$$\tilde{t}_{k+1}(\tau) = 2\tau\tilde{t}_k(\tau) + \tilde{t}_{k-1}(\tau),$$

we are able to construct real algorithms for inclusion ellipses with vertical major axis.
According to the fact

$$\tilde{t}_k(i\tau) = i^k t_k(\tau)$$

(see [MEYER 2]) we have

$$\frac{\tilde{t}_k(z_2)}{\tilde{t}_k(z_1)} \to 0 \qquad (k \to \infty),$$

if $z_1 \in i \cdot \mathbb{E}(\beta_1)$, $z_2 \in i \cdot \mathbb{E}(\beta_2)$ and

$$\beta_1 > \beta_2 .$$

(Here, i denotes the imaginary unit.)
Note that $i\mathbb{E}(\beta)$ represents an ellipse with focuses $+i$ and $-i$.
Hence, this polynomial can be applied for two special cases:

1. $\mathcal{A}$ is a skew-symmetric operator ($\mathcal{A} = -\mathcal{A}^*$) having pure imaginary eigenvalues. Then the iteration with $\tilde{t}_k(\alpha'\mathcal{A})$ yields absolutely largest eigenvalues of $\mathcal{A}$. The number α' should fulfil

$$|\lambda_{p+1}| \leq \alpha' < |\lambda_p| .$$

   Such an approach can be found in [DIETRICH 1].

2. If $\mathcal{A}$ is non-symmetric, but has Perron-Frobenius-properties represented by an non-negative matrix, the iteration with $\tilde{t}_k(\alpha'\mathcal{A})$ also leads to an accelerated power method for the eigenvalue $\lambda_1 = \rho(\mathcal{A})$. Indeed $\rho(\mathcal{A}) > 0$ is the eigenvalue on the boundary of the largest ellipse $i\mathbb{E}(\beta)$.

For both cases the real arithmetic is kept for real operators $\mathcal{A}$ using the recurrent definition of $\tilde{t}_k$.

3.6. Lanczos-Algorithms

Even though this volume is mainly devoted to other modern non-Lanczos procedures for large sparse eigenvalue problems, we will deal with the basic idea and some design features of the Lanczos-algorithms here for purposes of completeness. The present state of the design and use of the Lanczos-algorithms may be referred to in detail in [CULLUM/WILLOUGHBY 1] or [PARLETT 1,2].

The basic one-vector form of this algorithm was devised by [LANCZOS 1] for tridiagonalizing a symmetric matrix (more generally, a symmetric operator as used here), but due to the inherent instabilities this method remained unpopular among numerical analysts over the years. However, recent research (especially [PAIGE 1,2,3 et al]) and numerical experimentation have renewed the interest in this method.
In order to describe this algorithm we again provide $\mathcal{A}$ to be a symmetric operator mapping $\mathbf{H}$ into itself. In applications we will use the same operator as for Simultaneous Iteration or for the methods of 3.4. This will be discussed in Chapters 5.2. and 5.3.
The Lanczos-procedure is directed to produce an orthonormal basis $\tilde{Z}_s = (z_1,\ldots,z_s)$ of the Krylov-subspace

$$\mathbf{K}_{s-1}(\mathcal{A},z_1) = \text{span}\,(z_1,\ \mathcal{A}z_1,\ldots,\mathcal{A}^{s-1}z_1)$$

and consists in the following basic algorithm.

Basic Lanczos-Recursion (single vector version)

Let $\|z_1\| = 1, \quad \beta_1 = 0,$

for $j = 1,\ldots,s$ define

$$\beta_{j+1}\, z_{j+1} = \mathcal{A}z_j - \alpha_j z_j - \beta_j z_{j-1} \tag{3.6.1}$$

with $\alpha_j = \langle \mathcal{A}z_j, z_j\rangle$, (3.6.2)

$$\beta_{j+1} = \langle \mathcal{A}z_j, z_{j+1}\rangle \tag{3.6.3}$$

$$(= \|\mathcal{A}z_j - \alpha_j z_j - \beta_j z_{j-1}\|).$$

For each j the corresponding Lanczos-matrix T_j is the tridiagonal matrix with diagonal entries $\alpha_1, \alpha_2, \ldots, \alpha_j$

and sub-diagonal (super-diagonal) entries

$\beta_2, \ldots, \beta_j$.

Note that β_{j+1} has been calculated in defining T_j. From (3.6.1) - (3.6.3) we conclude for

$$\tilde{Z}_s = (z_1, \ldots, z_s) \in \mathbf{H}^s$$

$$\tilde{Z}_s^H \tilde{Z}_s = I \quad \text{and} \quad \tilde{Z}_s^H \mathcal{A} \tilde{Z}_s = T_s , \tag{3.6.4}$$

hence T_s is the matrix representation of $\mathcal{A}$ projected into $\mathbf{K}_{s-1}(\mathcal{A}, z_1)$, the Ritz-projection of $\mathcal{A}$.
From the fact that extremal eigenvalues of T_j converge rapidly to some extremal eigenvalues of $\mathcal{A}$ we can determine eigenvalue approximations for $\mathcal{A}$ by solving the eigenproblem of T_j (for some or each j).
Obviously, the Ritz-vectors x_j of span $\tilde{Z}_j$ belonging to these eigenvalues are eigenvector approximations.
From

$$\mathcal{A}\tilde{Z}_j = \tilde{Z}_j T_j + \beta_{j+1} z_{j+1} e_j^T \tag{3.6.5}$$

$$(e_j = (0, \ldots, 0, 1)^T \in \mathbf{R}^j)$$

the residuals for such Ritz-vectors are known before the Ritz-vectors are calculated:
If

$$T_j q_j = \delta_j q_j , \tag{3.6.6}$$

(one typical eigensolution for T_j)
then

$$x_j = \tilde{Z}_j q_j$$

fulfils:

$$\|r(x_j)\| = \|\mathcal{A}x_j - \delta_j x_j\|$$

$$= \|\mathcal{A}\tilde{Z}_j q_j - \delta_j \tilde{Z}_j q_j\|$$

$$= \beta_{j+1} |e_j^T q_j| \; . \qquad (3.6.7)$$

This will be used for the investigation into whether some eigenvalues are calculated with expected accuracy. The main problem of the implementation (in any application) results from the finite precision arithmetic. Practically, all the formulas mentioned above (especially (3.6.4)) are entirely destroyed after some Lanczos-steps, due to the cancellation in the recurrent calculations. Surprisingly, even if T_j is no longer a Ritz-projection matrix, it contains enough information about some good eigenvalue approximations. This observation is mainly used by [CULLUM/WILLOUGHBY 1] to propose Lanczos methods without any reorthogonalization processes. Here, the step number s will often become a large number (even larger than the dimension n). Furthermore, this method requires a practical device for the distinction of the true eigenvalues from the spurious ones.
If the computational effort for the calculation of the matrix-vector-product $\mathcal{A}z_j$ is high enough, procedures with reorthogonalization steps are superior, which ensure a minimal step size s (often in the order of $\sqrt{n}$). Such reorthogonalization techniques can be found in [PARLETT 1, MATTHIES 1, SIMON 1 and others].
In order to compute eigenvectors of multiple eigenvalues the block-Lanczos-method was proposed [UNDERWOOD 1, GOLUB/UNDERWOOD 1, CULLUM 1 et al] . Furthermore, the convergence characteristics are improved due to the simultaneous use of p columns together for the definition of a generalized Krylov-subspace

$$K_s(\mathcal{A}, Z_o) = \text{span}\,(Z_o \vdots \mathcal{A} Z_o \vdots \ldots \vdots \mathcal{A}^s Z_o)$$

for $Z_o \in H^p$.
This leads to a generalization of the formulas (3.6.1) - (3.6.3):

Block-Lanczos-Recursion:

Let $Z_1 \in \mathbb{H}^p$ with $Z_1^H Z_1 = I$,

$$B_1 = \mathbb{O} \in \mathbb{R}^{p \times p},$$

for $j = 1,\ldots,s$ define

$$Z_{j+1}B_{j+1} = \mathcal{A} Z_j - Z_j A_j - Z_{j-1}B_j^T \tag{3.6.8}$$

with $$A_j = Z_j^H \mathcal{A}\, Z_j, \tag{3.6.9}$$

$$B_{j+1} = Z_{j+1}^H \mathcal{A}\, Z_j \tag{3.6.10}$$

(or B_{j+1} guarantees $Z_{j+1}^H Z_{j+1} = I$).

Here, for each j the Lanczos-matrix $T_j \in \mathbb{R}^{pj \times pj}$ is block tridiagonal, symmetric with

diagonal blocks $A_1,\ldots,A_j$,

sub-diagonal blocks $B_2,\ldots,B_j$ and

super-diagonal blocks $B_2^T,\ldots,B_j^T$.

Let again

$$\tilde{Z}_s = (Z_1 \vdots Z_2 \vdots \ldots \vdots Z_s) \in \mathbb{H}^{p \cdot s},$$

then we have theoretically

$$\tilde{Z}_s^H \tilde{Z}_s = I, \quad \tilde{Z}_s^H \mathcal{A} \tilde{Z}_s = T_s.$$

There are two basic features for the use of the Lanczos-algorithm for eigenvalue calculation.

a) The Lanczos-recursion is performed until convergence of all desired eigenvalues. This could imply that the recursion is stopped for computing a Ritz-vector in order to orthogonalize all further Lanczos-vectors with respect to it (see [PARLETT 1]).

b) We choose (at the beginning or during the Lanczos-run) a small number s for the maximal dimension of the Krylov-subspace. Then the desired number of the algebraically largest eigenvalues and eigenvectors of

T_s are calculated and with the Ritz-vectors obtained from these eigenvectors the process is repeated. Especially for the block-Lanczos-procedure this iterative fashion is often used. If s is fixed at the beginning, we obtain the simultaneous generalization of the s-step method of steepest ascent (according to [FADDEYEV/FADDEYEVA 1]), that is $\mu(\hat{X})$ is maximized over the subspace $\mathbf{K}_{s-1}(\mathcal{A}, Z_1)$. Here, $\hat{X}$ denotes the calculated Ritz-tuple after the Lanczos-process started with $X = Z_1$:

$$T_s Q = Q \hat{\Delta} ,$$

$$Q \in \mathbb{R}^{p \cdot s \times p}, \ \hat{\Delta} = \operatorname{diag} (\hat{\delta}_1, \ldots, \hat{\delta}_p),$$

$$\hat{X} = \tilde{Z}_s Q.$$

Obviously, the method of steepest ascent in 3.4. is the special case for $s = 2$.

All in all, the design of a convenient Lanczos-algorithm for a given problem depends on various parameters: the computational costs of $y := \mathcal{A}x$ with respect to the costs of $z := \alpha x + \beta y$, the number of desired eigenvalues, the distribution of the spectrum of $\mathcal{A}$ (large gaps imply quick convergence). Furthermore, the spectral transformation techniques of Chapter 5.3. could also be applied here (cf. [ERICSSON/RUHE 1]), that is, the use of a function $\mathcal{C} = f(\mathcal{A})$ instead of $\mathcal{A}$.

4. CONVERGENCE PROPERTIES

In this chapter the questions of convergence of the algorithms stated at Chapter 3 are investigated. We prove the important basic property that simultaneous methods of steepest ascent and its accelerations of Chapter 3.2. and 3.3. are convergent, yielding some p linear independent eigensolutions of the problem $\{\mathcal{A}, \mathcal{B}\}$. Furthermore, for the method of steepest ascent the rate of convergence is estimated.

4.1. Convergence to Eigenelements

Here, we prove the convergence of the columns of X to p **linearly independent** eigenvectors of $\{\mathcal{A}, \mathcal{B}\}$. This result will be a direct consequence of the following lemma considering the basic algorithm of Chapter 3.1.

Lemma 4.1.1. One step of the iteration of the basic algorithm (3.1.1) - (3.1.4) using X and S and yielding $\hat{X}$ satisfies the inequality

$$\|S^H R\|_F^2 \leq \varkappa_0 \, (\lambda_1 - \lambda_n) \, \|S^H \mathcal{B} \, S\|_2 \quad (\mu(\hat{X}) - \mu(X))$$

with $\quad \varkappa_0 = \frac{1}{2}(1 + \sqrt{2})$

and $\quad R = \mathcal{A} X - \mathcal{B} X \Delta .$

Proof. 1. For the beginning, let us consider the case $\mathcal{B} = \mathcal{I}$ (identity operator) and in addition to (3.1.4) let S be the Ritz-basis of

$$\text{span } S \perp \text{span } X,$$

that is

$$X^H X = I, \; S^H S = I, \; X^H S = \mathbb{O}, \tag{4.1.1}$$

$$X^H \mathcal{A} X = \Delta = \text{diag } (\delta_1, \ldots, \delta_p),$$

$$S^H \mathcal{A} S = \Sigma = \text{diag } (\sigma_1, \ldots, \sigma_m) \tag{4.1.2}$$

with

$$\delta_1 \geq \ldots \geq \delta_p, \tag{4.1.3}$$

$$\sigma_1 \geq \ldots \geq \sigma_m. \tag{4.1.4}$$

The p-tuple $\hat{X}$ was constructed to have maximal value $\mu(\hat{X})$ over the subspace span $(X \vdots S)$, thus, if

$$X' = X + SU,$$

we have $\mu(\hat{X}) \geq \mu(X')$

or

$$\mu(\hat{X}) - \mu(X) \geq \mu(X') - \mu(X)$$

for any (mxp)-matrix U.
Using the definition (2.4.1), the difference $\mu(X') - \mu(X)$ becomes

$$\mu(X') - \mu(X) = \mathrm{tr}(X'^H X')^{-1}(X'^H \mathcal{A} X') - \mathrm{tr}\,\Delta$$

$$= \mathrm{tr}(I+U^TU)^{-1}(\Delta + U^T\Sigma U + U^TS^H\mathcal{A} X + X^H\mathcal{A} SU) - \mathrm{tr}\,\Delta$$

$$= \mathrm{tr}(I+U^TU)^{-1}(U^T\Sigma U - U^TU\Delta + U^TS^H\mathcal{A} X + (U^TS^H\mathcal{A} X)^T) \quad .$$

Together with Lemma 2.1.2 and the fact

$$S^H\mathcal{A} X = S^H(\mathcal{A} X - X\Delta) = S^HR$$

this leads to the inequality

$$\mu(\hat{X}) - \mu(X) \geq \mathrm{tr}(I+U^TU)^{-1}(U^T\Sigma U - U^TU\Delta + 2U^TS^HR)$$

$$\geq \mathrm{tr}(I+U^TU)^{-1}(\sigma_m U^TU - \delta_1 U^TU + 2U^TS^HR). \tag{4.1.5}$$

(For the second term of (4.1.5) the matrix C of Lemma 2.1.3 was set to the symmetric matrix $(I+U^TU)^{-1}U^TU$).

If $U = \frac{1}{\alpha} S^HR$,

this inequality yields

$$\mu(\hat{X}) - \mu(X) \geq (\sigma_m - \delta_1 + 2\alpha)\ \mathrm{tr}(\alpha^2 I + (S^HR)^T(S^HR))^{-1}(S^HR)^T(S^HR)$$

for an arbitrary positive number α. If

$$\alpha = \frac{1}{2}\left(|\sigma_m - \delta_1| + \beta\right) \quad \text{with} \quad \beta > 0,$$

then $\sigma_m - \delta_1 + 2\alpha \geq \beta$

and $\alpha \leq \frac{1}{2}(\lambda_1 - \lambda_n + \beta)$.

Hence, the same trace-lemma further yields

$$\mu(\hat{X}) - \mu(X) \geq \beta\cdot \mathrm{tr}(S^HR)^TS^HR \cdot (\alpha^2 + \|S^HR\|_2^2)^{-1},$$

respectively

$$\|S^H R\|_F^2 \leq \frac{1}{\beta}(\alpha^2 + \|S^H R\|_2^2)\,(\mu(\hat{X}) - \mu(X)) \leq$$

$$\frac{1}{4\beta}\,((\lambda_1 - \lambda_n + \beta)^2 + 4\|S^H R\|_2^2)\,(\mu(\hat{X}) - \mu(X)). \qquad (4.1.6)$$

A general bound for $\|S^H R\|_2$ is $\frac{\lambda_1 - \lambda_n}{2}$ from the following estimation:

$$\|S^H R\|_2^2 < \|S^H S\|_2\,\|R^H R\|_2 = \|R^H R\|_2 = \max \|Rt\|^2 \quad \left(\|t\|_2 = 1\;,\; t \in \mathbb{R}^p\right).$$

The residual R is also the term

$$R = (\mathcal{I} - \mathcal{P}_x)\mathcal{A}X$$

with the orthogonal projection $\mathcal{P}_x$ onto span X (compare [VAN DER GRAFT 1]), but $(\mathcal{I} - \mathcal{P}_x)X = 0$, thus R also becomes

$$R = (\mathcal{I} - \mathcal{P}_x)(\mathcal{A} + \tau\mathcal{I})X \quad \text{for all} \quad \tau \in \mathbb{R}.$$

Hence,

$$\|S^H R\|_2^2 \leq \max_{\|t\|_2 = 1} \|(\mathcal{I} - \mathcal{P}_x)(\mathcal{A} + \tau)Xt\|^2 \leq \|\mathcal{A} + \tau\mathcal{I}\|^2$$

and by minimization over τ

$$\|S^H R\|_2^2 \leq \left(\frac{\lambda_1 - \lambda_n}{2}\right)^2.$$

This bounds the constants of the right hand side of (4.1.6) by

$$\|S^H R\|_F^2 \leq \frac{1}{4\beta}\,((\lambda_1 - \lambda_n + \beta)^2 + (\lambda_1 - \lambda_n)^2)\cdot(\mu(\hat{X}) - \mu(X))$$

and a second minimization over $\beta > 0$ completes the first part of the proof for $\mathcal{B} = \mathcal{I}$ and S with (4.1.1):

$$\|S^H R\|_F^2 \leq \varkappa_0(\lambda_1 - \lambda_n)(\mu(\hat{X}) - \mu(X)). \qquad (4.1.7)$$

2. For arbitrary directions of movement S we construct another basis $(X \vdots \tilde{S})$ of span $(X \vdots S)$, which fulfils (4.1.1)/(4.1.2). If $\mathcal{P}_x$ is again the orthogonal projection operator onto span X, then

$$S_1 = (\mathcal{I} - \mathcal{P}_x)S$$

has columns orthogonal to X, which are linearly independent due to the assumption dim span $(X \vdots S) = p + m$ for the basic algorithm. Thus $S_2 = S_1(S_1^H S_1)^{-\frac{1}{2}}$ satisfies (4.1.1) and the orthogonal matrix $\tilde{U}$ of all eigenvectors of $S_2^H \mathcal{A} S_2$ transforms S_2 into $\tilde{S} = S_2\tilde{U}$ with (4.1.2). From (4.1.7) ($\tilde{S}$ instead of S) we have from the unitary invariance of the Frobenius norm

$$\|(S_1^H S_1)^{-\frac{1}{2}} S_1 R\|_F^2 \leq \varkappa_0 (\lambda_1-\lambda_n)(\mu(\hat{X})-\mu(X)).$$

The trace-lemma again completes the proof, using the facts:

$$tr(S_1^H R)^T (S_1^H S_1)^{-1} S_1^H R \geq \|S_1^H R\|_F^2 / \|S_1^H S_1\|_2,$$

$$\|S_1^H S_1\|_2 = \max_{\|t\|_2=1} \|S_1 t\|^2 = \max \|(\mathcal{I}-P_X) St\|^2 \leq$$

$$\leq \|S^H S\|_2$$

and $S_1^H R = S^H R$ (because $R^H X = \mathbf{0}$).

Hence, if $\mathcal{B} = \mathcal{I}$, then

$$\|S^H R\|_F^2 \leq \varkappa_0 (\lambda_1-\lambda_n) \|S^H S\|_2 \ (\mu(\hat{X})-\mu(X)). \tag{4.1.8}$$

3. For arbitrary $\mathcal{B}$ (symmetric, positive definite) we transform the eigenvalue problem $\{\mathcal{A}, \mathcal{B}\}$ in $\mathbf{H}$ into the eigenvalue problem for the operator $\mathcal{C} = \mathcal{B}^{-1}\mathcal{A}$ in $\mathbf{H}_{\mathcal{B}}$, the same space with $\mathcal{B}$-inner product. The operator $\mathcal{C}$ is a symmetric operator due to the $\mathcal{B}$-inner product, thus for the basic algorithm for $\mathcal{C}$, X and the given direction S the result (4.1.8) is true, if instead of R a new residual $\tilde{R}$ of the operator $\mathcal{C}$,

$$\tilde{R} = \mathcal{C}X - X\Delta$$

$$= \mathcal{B}^{-1}(\mathcal{A}X - \mathcal{B}X\Delta)$$

$$= \mathcal{B}^{-1}R$$

is used.

With $\mathcal{B}$-inner product (4.1.8) yields now

$$\|S^H \mathcal{B} \tilde{R}\|_F^2 = \|S^H R\|_F^2 \leq \varkappa_0 (\lambda_1-\lambda_n) \ \|S^H \mathcal{B} S\|_2 \ (\mu(\hat{X})-\mu(X))$$

and the lemma is verified. ■

With the help of the previous lemma we are able to prove the convergence of the columns of X to p eigenvectors of $\{\mathcal{A}, \mathcal{B}\}$ for the algorithms of Chapters 3.1. and 3.2. by verifying $R(X) \to 0$. The following lemma shows the equivalence of this fact with the convergence to eigenpairs.

Lemma 4.1.2. If an algorithm produces the series $\{X_k\}$ and $\{\Delta_k\}$, $k = 1,2,\ldots,$ with $\Delta_k = X_k^H \mathcal{A} X_k$, $X_k^H \mathcal{B} X_k = I$ and

$$\Delta_{k+1} \geq \Delta_k, \tag{4.1.9}$$

then the convergence of $R_k = R(X_k) = \mathcal{A} X_k - \mathcal{B} X_k \Delta_k$ to zero is equivalent to the convergence of the columns of X_k to p linearly independent eigenvectors of $\{\mathcal{A}, \mathcal{B}\}$.

Proof. Let $V_k = \Phi^H \mathcal{B} X_k \in \mathbb{R}^{n\times p}$,
that is

$$X_k = \Phi V_k,$$

then $\quad R_k = \mathcal{B} \Phi (\Lambda V_k - V_k \Delta_k) \longrightarrow 0$
means

$$\Lambda V_k - V_k \Delta_k \longrightarrow \mathbb{O}. \tag{4.1.10}$$

The matrices V_k contain orthonormal columns

$$(X_k^H \mathcal{B} X_k = V_k^T V_k = I).$$

Hence, if the eigenvalues in Λ are distinct, (4.1.10) together with (4.1.9) is only possible, if the following convergence is valid

$$\Delta_k \longrightarrow \tilde{I} \Lambda \tilde{I} \text{ and } V_k \longrightarrow \tilde{I}.$$

Here, $\tilde{I}$ denotes a matrix containing p different coordinate unit vectors

$$\tilde{I} = (e_{j_1} \quad e_{j_2}, \ldots, e_{j_p}).$$

Thus $\quad X_k \longrightarrow \Phi \tilde{I} = \tilde{\Phi} = (\varphi_{j_1}, \ldots, \varphi_{j_p}) \in \mathbb{R}^p.$

The case of multiple eigenvalues is more complicated. If, for example, two diagonal elements of Δ_k ($\delta_1^{(k)}$, $\delta_2^{(k)}$ say) converge to an eigenvalue of multiplicity 3 (to $\lambda_1 = \lambda_2 = \lambda_3 > \lambda_4$ say), the first two columns of V_k need not converge at all, but all entries except the first three of these columns must tend to zero, whenever (4.1.10) is valid. Thus, in $X_k = \Phi V_k$ the first two columns need not converge, but are arbitrary good eigenvector approximations for $k \to \infty$ and linearly independent. The eigenvalue approximations in Δ_k are convergent to eigenvalues in any case. ∎

Now, the main theorems can be stated to justify the algorithms of Chapter 3. That is, with the help of Lemma 4.1.1 we verify the convergence of R to zero. Thus, from Lemma 4.1.2 these algorithms certainly yield p distinct eigenpairs.

Theorem 4.1.1. For the simultaneous algorithm of steepest ascent (gradient method, i.e. S = R) the residual R converges to zero.

Proof. With S = R and

$$\|S^H \mathcal{B} S\|_2 < \|\mathcal{B}\| \cdot \|R^H R\|_F$$

the Lemma 4.1.1. yields

$$\|R^H R\|_F \leq \varkappa_o(\lambda_1 - \lambda_n) \|\mathcal{B}\| (\mu(\hat{X}) - \mu(X)),$$

and the difference $\mu(\hat{X}) - \mu(X)$ certainly tends to zero, because $\mu(X)$ is bounded by $\mathrm{tr}\Lambda_1$. ■

Theorem 4.1.2. For the simultaneous conjugate gradient algorithm (see Ch. 3.2., S = R + WT) the residual R converges to zero.

Proof. From Chapter 3.2. we remember

$$W^H R = \mathbf{O},$$

thus $S^H R = R^H R$.

It remains to estimate the norm $\|S^H \mathcal{B} S\|_2$ for using Lemma 4.1.1. This was the result of Lemma 3.2.1:

$$\|S^H \mathcal{B} S\|_2 < \|R^H R\|_F \|\mathcal{B}\| (1 + \tfrac{p}{\gamma}(\lambda_1 - \lambda_n))^2$$

with a small fixed constant γ. Hence, Lemma 4.1.1 again yields

$$\|R^H R\|_F < \mathrm{const} \cdot (\mu(\hat{X}) - \mu(X)).$$

■

The proofs of these theorems cause some more general possibilities of choosing the directions of movement S, for which also convergence to eigenpairs is obtained. Such directions are

$$S = \mathcal{T} R \qquad (4.1.11)$$

for positive definite operators $\mathcal{T}$ possibly changing from step to step over the iteration, or even more general

$$S = (\mathcal{T}_1 r_1, \mathcal{T}_2 r_2, \ldots, \mathcal{T}_p r_p)' \qquad (4.1.12)$$

for positive definite operators $\mathcal{T}_i$ (also allowed to vary during the iteration process). The algorithms obtained have not been considered up to now, but such algorithms could be on further interest for varying preconditioning operators.

Theorem 4.1.3. For the basic algorithm using S according to (4.1.12) ((4.1.11) is a special case) the residual R converges to zero, if the symmetric operators $\mathcal{T}_i$ (possibly changing over the iteration process) are uniformly bounded by two fixed constants β_1 and β_2,

$$0 < \beta_1 \langle x,x\rangle \leq \langle \mathcal{T}_i x,x\rangle \leq \beta_2 \langle x,x\rangle \qquad (\forall x \in \mathbb{H}). \tag{4.1.13}$$

Proof. In order to use Lemma 4.1.1 we need the estimations

$$\| S^H R \|_F^2 = \sum_{i,j} \langle \mathcal{T}_i r_i, r_j \rangle^2 \geq \sum_{i=1}^{p} \langle \mathcal{T}_i r_i, r_i \rangle^2 \geq \frac{1}{p} \left(\sum_{i=1}^{p} \langle \mathcal{T}_i r_i, r_i \rangle \right)^2$$

$$\| S^H \mathcal{B} S \|_2 \leq \operatorname{tr} S^H \mathcal{B} S = \sum_{i=1}^{p} \langle \mathcal{B} \mathcal{T}_i r_i, \mathcal{T}_i r_i \rangle \leq \sum \langle \mathcal{T}_i \mathcal{T}_i^{\frac{1}{2}} r_i, \mathcal{T}_i^{\frac{1}{2}} r_i \rangle \|\mathcal{B}\| \leq$$

$$\leq \|\mathcal{B}\| \beta_2 \sum_{i=1}^{p} \langle \mathcal{T}_i r_i, r_i \rangle .$$

Hence, the lemma yields

$$\sum \langle \mathcal{T}_i r_i, r_i \rangle \leq p \|\mathcal{B}\| \beta_2 \varkappa_0 (\lambda_1 - \lambda_n) (\mu(\hat{X}) - \mu(X))$$

or

$$\operatorname{tr} R^H R = \| R \|^2 \leq p \|\mathcal{B}\| \frac{\beta_2}{\beta_1} \varkappa_0 (\lambda_1 - \lambda_n)(\mu(\hat{X}) - \mu(X)). \blacksquare$$

For the W-accelerated method of steepest ascent of Chapter 3.3. a corresponding convergence theorem can be stated:

Theorem 4.1.4. For the W-accelerated method of steepest ascent according to Chapter 3.3. the residual R converges to zero.

Proof. Here, the direction of movement S consists in the 2p-tuple $S = (R \vdots W)$ and it is difficult to obtain an estimation for $\| S^H \mathcal{B} S \|_2$. Hence, this proof has to start analogous to Lemma 4.1.1.
The algorithm calculates

$$\hat{X} = (X \vdots R \vdots W) Q$$

with maximal $\mu(\hat{X})$ over that subspace

$$\operatorname{span} (X \vdots R \vdots W).$$

Let $X' = (X \vdots R) U = (X \vdots R \vdots W) \begin{pmatrix} U \\ 0 \end{pmatrix}$, $U \in \mathbb{R}^{2p \times p}$,

then, obviously,

$$\mu(\hat{X}) \geq \mu(X'). \qquad (4.1.14)$$

If U is chosen to form the Ritz-tuple X' with maximal value $\mu(X')$ over the subspace span $(X\vdots R)$, then X' represents the next iterated element of the simple method of steepest ascent (starting with the same element X). Hence Lemma 4.1.1 yields

$$\|R^H R\|_F \leq \varkappa_o\ (\lambda_1 - \lambda_n)\ \|\mathcal{B}\|\,(\mu(X') - \mu(X))$$

(according to the proof of Theorem 4.1.1).
The inequality (4.1.14) completes the proof and for the W-accelerated method the same estimation

$$\|R^H R\|_F \leq \varkappa_o\ (\lambda_1 - \lambda_n)\ \|\mathcal{B}\|\,(\mu(\hat{X}) - \mu(X))$$

is valid. ∎
Summarizing the results of these theorems, it is guaranteed that all algorithms described in Chapters 3.1. - 3.4., which are based on Ritz-steps for maximizing $\mu(X)$, will produce arbitrary exact eigenpair approximations. The eigenvalue approximations generated by these algorithms are monotonically non-decreasing and converge to p eigenvalues $\lambda_{j_1}, \ldots, \lambda_{j_p}$ of the original problem $\{\mathcal{A}, \mathcal{B}\}$.
All practical experiments, however, create only one result: the first p eigenvalues $\lambda_1, \ldots, \lambda_p$ are obtained ($\mu(X)$ is maximized indeed). Unfortunately, no general proof of this fact has been known up to now. For the special case $\mathcal{B} = \mathcal{I}$ and p = 1 the proof of convergence of the gradient method to λ_1 can be found in [FADDEYEV/FADDEYEVA 1]. [LONGSINE/McCORMICK 1] described a similar simultaneous procedure (i.e. $p > 1$) with a gradient direction of movement (S=R), but without total maximization of $\mu(\hat{X})$ at every step. Here, the fact was proven that, if $\mathcal{B} = \mathcal{I}$, then φ_1 is contained in the limiting subspace, thus

$$\lambda_{j_1} = \lambda_1$$

for this algorithm.
If for the eigenvalue approximations $\delta_1 \geq \ldots \geq \delta_p$ at one step the inequality

$$\delta_p \geq \lambda_{p+1} \qquad (4.1.15)$$

is satisfied, obviously the convergence of Δ to Λ_1 is guaranteed, but for practical purposes it is meaningless to use (4.1.15) as an assumption for the algorithm (compare [HANZAL 1]).

On the other hand, no counterexample is known to disturb this "ordered" convergence ([LONGSINE/McCORMICK 1]) of Δ to Λ_1 except for the case $\mathcal{B} = \mathcal{I}$, if the starting tuple X_o is deficient in some first eigendirections.

For $\mathcal{B} = \mathcal{I}$ at the k-th step of the gradient method (or its accelerations) every column of $X = X_k$ can be written as

$$x_i = \sum_{j=1}^{p} p_{ij}(\mathcal{A})\, x_j^{(o)} \tag{4.1.16}$$

with certain polynomials $p_{ij}(\tau)$ of degree k. Hence, entire presence of span Φ_1 in the starting subspace span X_o becomes necessary for the convergence of span X to spanΦ_1. This is equivalent to the assumption that the (pxp)-matrix

$$H_o = \Phi_1^H X_o \quad \text{is non-singular} \tag{4.1.17}$$

as is well-known for the Simultaneous Iteration Method (Chapter 3.5.) in order to ensure convergence to p absolutely largest eigenvalues of $\mathcal{A}$.

For the more complicated case $\mathcal{B} \neq \mathcal{I}$ a representation of X_k according to (4.1.16) becomes false and (4.1.17) does not seem to be necessary for the convergence. Here, if

$$X_o = \Phi_1 H_o + \Phi_2 G_o,$$

the first direction of movement is

$$R_o = \mathcal{A} X_o - \mathcal{B} X_o \Delta_o = \mathcal{B}\Phi_1(\Lambda_1 H_o - H_o \Delta_o) + \mathcal{B}\Phi_2(\Lambda_2 G_o - G_o \Delta_o)$$

and might contain all eigendirections $\varphi_1, \dots, \varphi_p$ even if H_o has zero rows! This depends on the eigenexpansion of the basis $\mathcal{B}\Phi$:

$$\mathcal{B}\Phi = \Phi\, \Theta \tag{4.1.18}$$

Here, Θ denotes the (nxn)-matrix with the coefficients θ_{ij} from

$$\mathcal{B}\varphi_i = \sum_{j=1}^{n} \theta_{ji}\, \varphi_j .$$

According to (4.1.18), Θ is also the matrix representation of the operator $\mathcal{B}$ with respect to the basis Φ. Thus the formulas

$$\Theta = \Phi^H \mathcal{B}^2 \Phi = (\Phi^H \Phi)^{-1} \tag{4.1.19}$$

are direct consequences of (4.1.18).

If $\mathcal{B} = \mathcal{I}$, then $\Theta = I$ and, if $\mathcal{A}$ and $\mathcal{B}$ commute, Θ still becomes a diagonal matrix, because of the orthogonality of the eigenvectors for this case. In general Θ represents a full dense matrix, thus

$$R_o = \mathcal{B}\Phi\left(\Lambda\begin{pmatrix}H_o\\G_o\end{pmatrix} - \begin{pmatrix}H_o\\G_o\end{pmatrix}\Delta_o\right)$$

$$= \Phi\Theta\left(\Lambda\begin{pmatrix}H_o\\G_o\end{pmatrix} - \begin{pmatrix}H_o\\G_o\end{pmatrix}\Delta_o\right)$$

could contain all eigendirections even if X_o does not.
The matrix Θ is also successfully used for the investigation of the rate of convergence in the next chapter.

4.2. The Rate of Convergence of the Method of Steepest Ascent

This chapter presents estimates for the asymptotic rate of convergence of the eigenvalue approximations, if "ordered" convergence, that is

$$\Delta \longrightarrow \Lambda_1,$$

is assumed. If $\mathcal{B} \neq \mathcal{I}$, such estimates are known for the gradient method (S=R) only.
The direct investigation of this algorithm is complicated, but it helps to make a comparison with another similar algorithm considered by [LONGSINE/McCORMICK 1 , algorithm SIRQIT-G]. From the given eigenvector approximations in X (with $X^H\mathcal{B}X = I$, $X^H\mathcal{A}X = \Delta$) this algorithm constructs the next iterated element $\hat{X}_{SIRQIT}$ with its eigenvalue approximations $\hat{\Delta}_{SIRQIT}$ by using the direction of movement R(X) together with some $\mathcal{B}$-orthogonalization processes. Altogether a formula

$$\hat{X}_{SIRQIT} = XU_1 + RU_2$$

is fulfilled, but the matrices U_1 and U_2 consist in some procedures of maximizing column-Rayleigh-quotients and orthogonalizations. For details see [MEYER 3]. For our algorithm

$$\hat{X} = (X \vdots R)Q = XQ_1 + RQ_2$$

is satisfied, where the functional $\mu(\hat{X})$ and thus all values $\hat{\delta}_i$ are maximized over $Q_{1/2}$. Hence, from Chapter 2.4. we conclude

$$\hat{\Delta} > \hat{\Delta}_{SIRQIT}. \qquad (4.2.1)$$

Now, [LONGSINE/McCORMICK 1] gave an asymptotic bound for the quotient

$$\frac{\lambda_j - \hat{\delta}_{j\ SIRQIT}}{\lambda_j - \delta_j}.$$

By (4.2.1) this is also a bound for

$$\frac{\lambda_j - \hat{\delta}_j}{\lambda_j - \delta_j}$$

for our algorithm. Asymptotically, both algorithms show the same behaviour, that is, the bound is equally valid for both algorithms (and equally sharp).

The advantage of LONGSINE/McCORMICK's version consists in a theoretical simplification, which enables us to estimate the rate of convergence. For practical purposes, there is no further advantage, especially the necessary computational costs are higher than for our algorithm.

Before stating the main theorem, we have to define some small eigenvalue problems, following [LONGSINE/McCORMICK 1].

Let

$$\Phi_{1,j} = (\varphi_j, \varphi_{j+1}, \ldots, \varphi_p) \in \mathbb{R}^{p-j+1},$$

$$\Lambda_{1,j} = \mathrm{diag}\,(\lambda_j, \ldots, \lambda_p),$$

$$C_j = (\lambda_j I - \Lambda_2)^{\frac{1}{2}}\, \Phi_2^H B^2 \Phi_2\, (\lambda_j I - \Lambda_2)^{\frac{1}{2}} \in \mathbb{R}^{(n-p)\times(n-p)},$$

and

$$D_j = (\lambda_j I - \Lambda_2)^{\frac{1}{2}}\, \Phi_2^H B^2 \Phi_{1,j}\, (\lambda_j I - \Lambda_{1,j}) \cdot$$

$$\cdot \Phi_{1,j}^H B^2 \Phi_2\, (\lambda_j I - \Lambda_2)^{\frac{1}{2}} \in \mathbb{R}^{(n-p)x(n-p)}$$

then denote by $\overline{\gamma}_j$ the largest and by $\underline{\gamma}_j$ the smallest eigenvalue of C_j. Furthermore, the matrix pencil $\{D_j, C_j^2\}$ has the largest eigenvalue ρ_j. Note that the eigenvalues of $\{\mathcal{A}, \mathcal{B}\}$ and some submatrices of Θ define the numbers $\overline{\gamma}_j$, $\underline{\gamma}_j$ and ρ_j $(j=1,\ldots,p)$.

Theorem 4.2.1. With the above definitions the asymptotic rate of convergence for the simultaneous method of steepest ascent can be estimated as

$$(\lambda_j - \hat{\delta}_j) \leqslant \eta_j(\lambda_j - \delta_j), \quad j = 1,\dots,p$$

with the convergence quotients

$$\eta_j = 1 - (1-\xi_j)(1-\zeta_j)$$

and

$$\xi_j = \left(\frac{\bar{\gamma}_j - \underline{\gamma}_j}{\bar{\gamma}_j + \underline{\gamma}_j}\right)^2,$$

$$\zeta_j = \frac{\rho_j}{1 + \rho_j} .$$

Proof. [LONGSINE/McCORMICK 1] proved this estimate for the case of real symmetric matrices $\mathcal{A}$ and $\mathcal{B}$. Furthermore, the calculation of p smallest eigenvalues was considered. The generalization to symmetric operators $\mathcal{A}$ and $\mathcal{B}$ (with respect to the inner product $\langle .,. \rangle$) as well as the change in the order of the eigenvalues become straight forward. Thus, all details will be found in this paper.

Corollary 4.2.1. The important special case $\mathcal{B} = \mathcal{J}$ leads to $\zeta_j = 0$ and

$$\eta_j = \xi_j = \left(\frac{\lambda_j - \lambda_n - (\lambda_j - \lambda_{p+1})}{\lambda_j - \lambda_n + (\lambda_j - \lambda_{p+1})}\right)^2 = \left(\frac{1 - \gamma_j}{1 + \gamma_j}\right)^2 \tag{4.2.2}$$

with the "gap-size" (similar to [PARLETT 1])

$$\gamma_j = \frac{\lambda_j - \lambda_{p+1}}{\lambda_j - \lambda_n}, \quad j = 1,\dots,p. \tag{4.2.3}$$

Note that the same convergence quotient η_j belongs to the Simultaneous Iteration Method (compare Chapter 3.5.), if the optimally shifted operator $\mathcal{A} + \tau\mathcal{J}$ with

$$\tau = -\frac{\lambda_{p+1} + \lambda_n}{2}$$

is used. Here, for the rate of convergence (of eigenvalue approximations) we have from Theorem 3.5.2

$$\eta_j = \left(\frac{\lambda_{p+1} + \tau}{\lambda_j + \tau} \right)^2 ,$$

which, for the above optimal τ, yields indeed

$$\eta_j = \left(\frac{1 - \gamma_j}{1 + \gamma_j} \right)^2 .$$

Hence, our gradient algorithm (for $\mathcal{B} = \mathcal{I}$) has, asymptotically, the same rate of convergence as best shifted Simultaneous Iteration, but without any knowledge required on information about the spectrum of $\mathcal{A}$. Numerical experiments demonstrate a much more rapid convergence during earlier iteration steps. Thus, the gradient method will be superior to Simultaneous Iteration Method, if the computational costs per step will not greatly exceed those of Simultaneous Iteration. This will be discussed in Chapter 5.

For practical considerations of the general case $\mathcal{B} \neq \mathcal{I}$ (more precisely: $\mathcal{A}\mathcal{B} \neq \mathcal{B}\mathcal{A}$) the estimates of Theorem 4.2.1 are less suitable, thus [LONGSINE/McCORMICK 1] gave bounds for ξ_j and ζ_j, which better relates their ratios to the eigenvalues of $\{\mathcal{A},\mathcal{B}\}$ and to the condition number $\varkappa(\mathcal{B})$ of the operator $\mathcal{B}$. This is

Theorem 4.2.2. The convergence quotients η_j of Theorem 4.2.1 can be estimated by

$$\eta_j = 1 - (1 - \xi_j)(1 - \zeta_j), \quad j = 1,\dots,p,$$

$$\xi_j \leqslant \left(\frac{1 - \xi\gamma_j}{1 + \xi\gamma_j} \right)^2, \tag{4.2.4}$$

$$\zeta_j \leqslant \frac{(1-\xi)^2(\lambda_j - \lambda_p)}{(1-\xi)^2(\lambda_j - \lambda_p) + 4\xi(\lambda_j - \lambda_{p+1})} \leqslant \left(\frac{1 - \xi}{1 + \xi} \right)^2 \tag{4.2.5}$$

with

$$\xi = (\varkappa(\mathcal{B}))^{-1} .$$

Corollary 4.2.2. For $\mathcal{B} \neq \mathcal{I}$ a convenient rate of convergence is guaranteed, if both the "gaps" γ_j and the inverse condition number ξ do not become too small.

4.3. Some Remarks to the Rate of Convergence of the Accelerated Gradient Methods

An entire theoretic result on the convergence of the Conjugate Gradient Method of Chapter 3.2. and of the W-Accelerated Gradient Method of Chapter 3.3. is unknown up to now. Therefore, we present in this Chapter some ideas, which are consequences of numerical experiments.

For purposes of simplicity we restrict ourselves to the case $\mathcal{B} = \mathcal{I}$ only. For this case some interesting connections between the W-Accelerated Method and the Simultaneous Chebyshev Iteration has been remarked in Chapter 3.3., furthermore a connection to the Lanczos algorithm can also be found. Here, we are going to compare the rate of convergence of these three algorithms. Because of using polynomial representations of the single iterated elements we will further restrict ourselves to the case p = 1 (iteration with one vector only for calculation of λ_1 and φ_1). The generalization to p > 1 is straightforward, however, one polynomial $p_k(\mathcal{A})$ of the operator $\mathcal{A}$ is to be replaced by a (pxp)-matrix of polynomials $(p_{ij}^{(k)}(\mathcal{A}))$ in order to represent the k-th iterated element X_k.

Here, we will consider three series of vectors:

y_k: generated by Lanczos' algorithm (the eigenvector approximation after k Lanczos steps),

x_k: generated by the W-accelerated gradient method,

z_k: generated by "best" Chebyshev accelerated power method.

To be more precise, we write down the formulas defining x_k, y_k and z_k. We denote by

$$\mathbb{K}_m(\mathcal{A}, x) = \text{span}\,(x, \mathcal{A}x, \ldots, \mathcal{A}^m x)$$

the Krylov-subspace of the operator $\mathcal{A}$ of dimension (m+1). All three methods use the same starting vector $x_0 = y_0 = z_0$. Hence,

(1) y_k is that vector in $\mathbb{K}_k(\mathcal{A}, x_0)$ with maximal Rayleigh quotient $\mu(y_k)$,

(2) x_k is that vector in span $(\mathcal{A}x_{k-1}, x_{k-1}, x_{k-2})$ with maximal Rayleigh quotient $\mu(x_k)$,

(3) $$z_k = (\alpha\mathcal{A}z_{k-1} + \beta z_{k-1} - \tau_{k-2}z_{k-2})\tau_{k-1} \qquad (4.3.1)$$

from Chapter 3.5.

Here, τ_k denote the normalizing factors in the Chebyshev iteration and α, β are (according to 3.5.)

$$\alpha = 2\alpha', \qquad \beta = 2\beta',$$

$$\alpha' = \frac{2}{\bar{\lambda} - \underline{\lambda}}, \quad \beta' = -\frac{\bar{\lambda} + \underline{\lambda}}{\bar{\lambda} - \underline{\lambda}}.$$

From the fact that $[\underline{\lambda}, \bar{\lambda}]$ contains all unwanted eigenvalues $\lambda_n, \ldots, \lambda_2$, we characterize the "best" Chebyshev accelerated power method by

$$\underline{\lambda} = \lambda_n \text{ and } \bar{\lambda} = \lambda_2.$$

Hence, $g(\tau) = \alpha'\tau + \beta'$ transforms the interval $[\lambda_n, \lambda_2]$ into $[-1,1]$ and (according to 3.5.)

$$z_k = \mathbf{t}_k(g(\mathcal{A}))x_0\tau_0 \ldots \tau_k \tag{4.3.2}$$

with the Chebyshev polynomial $\mathbf{t}_k(\tau)$ of degree k. From (1), (2) and (3) we conclude that all three vectors x_k, y_k, z_k are elements of $\mathbf{K}_k(\mathcal{A}, x_0)$.
Obviously,

$$\mu(x_k) \leqslant \mu(y_k), \tag{4.3.3}$$

$$\mu(z_k) \leqslant \mu(y_k). \tag{4.3.4}$$

From the first two steps of the iteration we have

$$x_1 = y_1 \quad \text{and} \quad x_2 = y_2,$$

thus

$$\mu(z_1) \leqslant \mu(x_1),$$

$$\mu(z_2) \leqslant \mu(x_2).$$

Numerical experiments show that the inequality

$$\mu(z_k) \leqslant \mu(x_k) \tag{4.3.5}$$

remains valid at the later steps also, but there is no proof up to now. If (4.3.5) would hold in general, then together with (4.3.3) we would have a two sided estimation of $\mu(x_k)$:

$$\mu(z_k) \leqslant \mu(x_k) \leqslant \mu(y_k). \tag{4.3.6}$$

The rate of convergence of $\mu(z_k)$ and of $\mu(y_k)$ is well-known from the literature and can be estimated with the same convergence quotient. According to [PARLETT 1] or [FADDEYEV/FADDEYEVA 1]

$$\lambda_1 - \mu(y_k) \leq \frac{\text{const}}{t_k^2(g(\lambda_1))} (\lambda_1 - \mu(x_0)).$$

An analogous estimation of $\lambda_1 - \mu(z_k)$ is concluded from Chapter 3.5. (in generalization of Theorem 3.5.2 for instance):

$$\lambda_1 - \mu(z_k) \leq \frac{\text{const}}{t_k^2(g(\lambda_1))} .$$

The denominator of these estimations can be transformed into the k-th power of a new convergence quotient (following [SAMARSKIJ1]):

$$g(\lambda_1) = \frac{2\lambda_1 - \lambda_2 - \lambda_n}{\lambda_2 - \lambda_n} = \frac{1 + \gamma_1}{1 - \gamma_1} ,$$

$$\gamma_1 = \frac{\lambda_1 - \lambda_2}{\lambda_1 - \lambda_n} ,$$

$$t_k\left(\frac{1 + \gamma_1}{1 - \gamma_1}\right)^{-1} = \zeta_k \left(\frac{1 - \sqrt{\gamma_1}}{1 + \sqrt{\gamma_1}}\right)^k$$

with $0 < \zeta_k < 2$.

Hence, the convergence of $\mu(z_k)$ or $\mu(y_k)$ to λ_1 is estimated as

$$\left.\begin{matrix} \lambda_1 - \mu(z_k) \\ \lambda_1 - \mu(y_k) \end{matrix}\right\} \leq (\eta_1')^{2k} \cdot \text{const}$$

with the better convergence quotient

$$\eta_1' = \left(\frac{1 - \sqrt{\gamma_1}}{1 + \sqrt{\gamma_1}}\right)^2 .$$

(The optimal shifted power method and the simple gradient method have the quotient

$$\eta_1 = \left(\frac{1 - \gamma_1}{1 + \gamma_1}\right)^2 .)$$

The numerical experiments suggest that for $(\lambda_1 - \mu(x_k))$ the same quotient η_1' is valid (a general proof of (4.3.6) would justify this).
These considerations shall illustrate the typical behaviour of the algorithms for the simultaneous case ($p > 1$) also. In this case the convergence quotients are replaced by

$$\left(\frac{1 - \sqrt{\gamma_j}}{1 + \sqrt{\gamma_j}}\right)^2, \qquad j = 1, \ldots, p,$$

with the "gap-sizes" defined previously at (4.2.3). Here, we should especially stress that at one hand for the Lanczos algorithm and for "best" Chebyshev iteration the same convergence has been estimated (if nothing about the distribution of eigenvalues in $[\lambda_n, \lambda_2]$ is known!) but on the other hand, practically, the Lanczos algorithm is much superior to Chebyshev iteration. The comparison between Chebyshev iteration and W-accelerated method of steepest ascent (or conjugate gradient method) causes the same result - that is, the unknown behaviour of these new methods is more similar to Lanczos' algorithm.
There are at least three slight advantages of these algorithms (compared with the Lanczos algorithm):

- The simultaneous calculation with p vectors causes no problems.
- The dangerous lost of orthogonalities during the construction of the Lanczos basis will not occur. At any step of the iteration the eigenvector approximations are present (X_k).
- For the implementation of the Lanczos algorithm only three vectors ($p=1$) or rectangular matrices ($p > 1$) are necessary in the inner storage, but all Lanczos vectors (or matrices Z_k) have to be stored until convergence (may be at back storage). Here, 3 vectors (rectangular matrices) are sufficient for the whole iteration process.

5. THE APPLICATION TO $K\varphi = \omega^2 M\varphi$

In this chapter we are going to demonstrate the application of the gradient algorithm and its accelerations of Chapter 3 to practical problems arising in vibration analysis of structures. We develop a general strategy for avoiding the very poor rate of convergence known from earlier experiments in connection with these or related algorithms ([SCHWARZ 2,3]).
A preferable rate of convergence for those algorithms is obtained from the introduction of convenient preconditionings. The full range of preconditionings, as it is known from the theory of iterative solutions of linear systems, will be allowed to apply to eigenvalue problems also. The second part of this chapter deals with the important special case $\mathcal{A} = K^{-1}M$, $\mathcal{B} = \mathcal{I}$ and the last one with spectral transformation ideas.

5.1. Algorithms for a Generalized Eigenvalue Problem Using Preconditionings

Firstly, let us consider the given problem

$$K\varphi = \omega^2 M\varphi \tag{5.1.1}$$

and its typical properties. Eigenvalue problems (5.1.1) especially arise from the analysis of eigenoscillations of structures, where the large sparse matrices K (stiffness-matrix) and M (mass-matrix) have been calculated by the methods of finite elements or finite differences.
The matrices K and M are symmetric and positive definite (the generalization to the case of one indefinite matrix is straightforward, as will be seen later). The p smallest eigenvalues

$$\omega_1^2 \leq \ldots \leq \omega_p^2$$

are to calculate together with the corresponding eigenvectors

$$\varphi_1, \ldots, \varphi_p.$$

The physical or technical signification of the ω_i consists in the eigenfrequencies and φ_i describe the eigenoscillation forms.
Let us denote by h the discretization parameter in the calculation of K and M from the discretization process. In order to construct an appropriate procedure for solving (5.1.1) we will relate the computational costs to h.

The following properties of the problem with respect to its dependence on h are well-known (compare [FRIED 2, KORNEEV/LANGER 1]):

(1) The dimension n and the number of non-zero entries of K and M are on the same order in h ($O(h^{-2})$ for square problems, $O(h^{-3})$ for 3-dimensional discretization).

(2) The lowest eigenvalues ω_1^2 are near fixed numbers ($O(1)$).

(3) The largest eigenvalue ω_n^2 tends to infinity for $h \to 0$.

(4) The condition number of M is finite, the condition number $\varkappa(K)$ tends to infinity for $h \to 0$ (for example $\varkappa(K) = O(h^{-2})$ for square problems).

An immediate consequence of the properties (2) and (3) is the very poor convergence of the simultaneous gradient method according to [SCHWARZ 2, 3, LONGSINE/McCORMICK 1, WALDVOGEL 1 and others], whenever

$$\mathcal{A} = K \quad \text{and} \quad \mathcal{B} = M$$

(with usual Euclidian inner product) was used to solve (5.1.1) by minimizing $\mu(X)$. In this case, gap-numbers γ_j' analogous to Theorem 4.2.2 are

$$\gamma_j' = \frac{\omega_{p+1}^2 - \omega_j^2}{\omega_n^2 - \omega_j^2} \longrightarrow 0 \text{ for } h \to 0. \tag{5.1.2}$$

Thus, for finer discretization the rate of convergence becomes arbitrarily slow.
Due to Theorem 4.2.2 gap-numbers not tending to zero with $h \to 0$ are necessary for convenient convergence results.
If we use the following substitutions

$$\mathcal{A} = C^{-1}M, \qquad \mathcal{B} = C^{-1}K, \tag{5.1.3}$$

then (5.1.1) is equivalent to

$$\mathcal{A}\varphi = \lambda \mathcal{B} \varphi \tag{5.1.4}$$

where

$$\lambda = \frac{1}{\omega^2}$$

and the largest eigenvalues of (5.1.4)

$$\lambda_1 = \frac{1}{\omega_1^2} \geqslant \ldots \geqslant \lambda_p = \frac{1}{\omega_p^2}$$

are to calculate.
The gap-numbers become now

$$\gamma_j = \frac{\lambda_j - \lambda_{p+1}}{\lambda_j - \lambda_n} = \frac{\omega_j^{-2} - \omega_{p+1}^{-2}}{\omega_j^{-2} - \omega_n^{-2}} \quad , \; j = 1,\ldots,p,$$

which are on the order O(1) in h, because $\lambda_n = \omega_n^{-2} \to 0$ for $h \to 0$.
Note that γ_j is connected with γ'_j from (5.1.2) by

$$\gamma_j = \gamma'_j \quad \frac{\omega_n^2}{\omega_{p+1}^2} \gg \gamma'_j \ .$$

Hence, for any problem (5.1.1), possibly not having the properties (1) - (4), the gap-sizes are enlarged by the above definitions.
In order to apply our algorithms of the Chapters 3.1. to 3.3. we need symmetric operators $\mathcal{A}$ and $\mathcal{B}$ with respect to a proper chosen inner product. Let C be a symmetric, positive definite matrix, then the definition

$$\mathcal{F} = C$$

for the inner product matrix, that is

$$\langle x,y \rangle \;\; = y^T C x, \quad x,y \in \mathbb{R}^n,$$

yields symmetricity for $\mathcal{A}$ and $\mathcal{B}$ and the results of Chapters 3 and 4 are applicable.
According to Theorem 4.2.2 a rate of convergence not depending on h (i.e. for $h \to 0$, the quotients η_j do not converge to 1) is guaranteed by the two facts:

- γ_j do not converge to zero for $h \to 0$,
- $\xi = (\, \varkappa(\mathcal{B}\,))^{-1}$ does not converge to zero for $h \to 0$.

While the first fact follows from the previous substitution (5.1.3), the second one makes a demand for a "good preconditioning" C.
Because $\mathcal{B}$ is a symmetric operator in $\mathbb{H}$, we have

$$\varkappa(\mathcal{B}\,) = \lambda_{max}(\mathcal{B}\,)/\;\lambda_{min}(\mathcal{B}\,).$$

The eigenvalues of $\mathcal{B} = C^{-1}K$ are the eigenvalues of the generalized symmetric problem $\{K,C\}$ also.

Thus,

$$\xi \geq \beta_1 / \beta_n,$$

if

$$\beta_1 \langle x,x \rangle \leq \langle \mathcal{B} x,x \rangle \leq \beta_n \langle x,x \rangle \qquad \forall x \in H$$

respectively,

$$\beta_1 \, x^T C x \leq x^T K x \leq \beta_n \, x^T C x \qquad \forall x \in \mathbb{R}^n. \tag{5.1.5}$$

Consequently, if there exist such constants β_1 and β_n (fulfilling (5.1.5), independent of h), the convergence quotients η_j are, also independent of h, less than unity.
That is, the number of iterations required for the calculation of the eigenvalues $\lambda_1, \ldots, \lambda_p$ (up to a given relative error ε) with the help of the simultaneous method of steepest ascent for $\mathcal{A}$ and $\mathcal{B}$ is bounded by a fixed constant, depending on the eigenvalue distribution and on β_1 / β_n but not depending on the discretization parameter h.
To be more precise, let k be the iteration index, $\delta_j^{(k)}$ the approximation to the eigenvalue λ_j and $\alpha_j^{(k)}$ its relative error

$$\alpha_j^{(k)} = \frac{\lambda_j - \delta_j^{(k)}}{\lambda_j} \quad ,$$

then according to Theorem 4.2.2

$$\alpha_j^{(k)} \leq \eta_j^k \; \alpha_j^{(o)} \; , \; j = 1, \ldots, p.$$

Now, if $\alpha_j^{(k)} \leq \varepsilon$ was expected, this means

$$\tilde{k} = \max_j \frac{\ln \varepsilon - \ln \alpha_j^{(o)}}{\ln \eta_j} = \max_j \frac{\ln \alpha_j^{(o)} + \ln \frac{1}{\varepsilon}}{\ln \frac{1}{\eta_j}} \tag{5.1.4}$$

is the largest number of necessary iteration steps, which does not depend on h, if η_j do not $(j=1, \ldots, p)$. It should be noted that the total number of iterations required is much less than $\tilde{k}$, because the bound of Theorem 4.2.2 was only asymptotical.
On the other hand, we stress that all basic operations within one step of the iteration have an operation count proportional to the number of unknowns, if the solution of linear systems with the preconditioning C could be performed in an appropriate manner. Thus, for such a preconditioning satisfying (5.1.5) the following algorithm is a so-called "optimal" or "quick" solver of the partial eigenproblem (having a total

operation count proportional to the number of unknowns).
The resulting algorithm for $\mathcal{A}$ and $\mathcal{B}$ in (5.1.3) using C-inner product reads as follows (an immediate consequence of the basic algorithm (3.1.1) - (3.1.5), note that $X^HY = X^TCY$ for the inner product defined now).

<u>Preconditioned Simultaneous Method of Steepest Ascent (PSGM(K,M,C))</u>

(1) $R := C^{-1}(MX-KX\Delta)$. (5.1.7)

(2) $A := \begin{pmatrix} \Delta & X^TMR \\ R^TMX & R^TMR \end{pmatrix}$,

$$B := \begin{pmatrix} I & X^TKR \\ R^TKX & R^TKR \end{pmatrix}. \qquad (5.1.8)$$

(3) Solve $AQ = BQ\hat{\Delta}$,

$\hat{\Delta} = \mathrm{diag}\,(\hat{\delta}_1, \ldots, \hat{\delta}_p)$, the p largest eigenvalues of $\{A,B\}$,

$Q^TBQ = I$.

(4) $\hat{X} := (X \vdots R)Q$. (5.1.9)

<u>Remark.</u> We should emphasize that every positive definite matrix C produces an algorithm of steepest ascent for $\mu(X)$ by producing the direction of steepest ascent R. The inner product space $\mathbf{H}$ and, hence, the 'steepest direction' is influenced by C. The functional $\mu(X)$, here, is independent of C, because

$$F(X) = (X^H\mathcal{B}\,X)^{-1}X^H\mathcal{A}\,X = (X^TCC^{-1}KX)^{-1}(X^TCC^{-1}MX) = (X^TKX)^{-1}(X^TMX)\ .$$

The use of preconditioning in methods for large sparse eigenvalue problems has also been investigated by [RUHE 1, 2, EVANS 1 or EVANS/SHANEHCHI 1] in connection with convergent splittings. The main advantage of our approach consists in the fact that all knowledge about preconditioning from the theory of iterative solution of linear systems is directly applicable to eigenvalue problems. A similar algorithm for the special case p = 1 has been described by [PRIKAZČIKOV 1].
With the special choice C = I the gradient methods in [SCHWARZ 1, 2, McCORMICK/LONGSINE 1, SAMEH/WISNIEWSKI 1 and others] are contained in PSGM.
In the theory of iterative solution of linear systems the construction of a so-called "spectral-equivalent" preconditioning C, that is

$$\beta_1\ x^TCx \leq x^TKx \leq \beta_n\ x^TCx$$

with constant β_n/β_1, often represents a difficult problem. On the other hand, a large number of preconditionings, having β_n/β_1 not constant but slightly growing with $h \to 0$, has been developed during the last ten years.
Especially, we have used the incomplete factorization methods for K to construct preconditionings C.
The main result using these preconditionings consists in the fact that the necessary number of iterations is essentially decreased in comparison with C = I. Thus, the prime result in [SCHWARZ 1, 2] that gradient methods are often less effective than coordinate relaxation methods, no longer remains true, if such simple and universal preconditionings as incomplete factorizations are used.
We will demonstrate the typical behaviour of PSGM by the following example:
Calculate some of the lowest eigenvalues of an elliptic operator with varying coefficients:

$$\mathcal{L} u = - p_1(x,y) \frac{\partial^2 u}{\partial x^2} - p_2(x,y) \frac{\partial^2 u}{\partial y^2} + p(x,y)u$$

for $(x,y) \in \Omega$
and $u = 0$ on the boundary of Ω.

Let $p_{1/2}$ and p positive smooth functions. Then 5-point-difference approximation leads to the large sparse matrix K and M = I (F.E.M. possibly yields another K and M $\neq$ I with the same typical properties of the eigenvalue problem, thus, here, we restrict ourselves to a 5-point-difference approximation). Let Ω be a rectangular domain, then a spectral equivalent preconditioning $C = C_0$ arises, if $\mathcal{L}$ is approximated by an operator $\tilde{\mathcal{L}}$ with fixed coefficients:

$$\tilde{\mathcal{L}} u = - q_1 \frac{\partial^2 u}{\partial x^2} - q_2 \frac{\partial^2 u}{\partial y^2} + qu.$$

The 5-point-discretization of $\tilde{\mathcal{L}}$ yields the matrix C_0 only depending on the constants $q_{1/2}$, q and the mesh size h, for which quick solvers exist (e.g. Fast Fourier Transform using the well-known eigenvalues and eigenvectors of such matrices). For a proper approximation of $\mathcal{L}$ we will use the averages of $p_{1,2}$ and p over Ω for the definition of $q_{1/2}$ and q, resp.

$$q_1 = \frac{1}{\text{mes } \Omega} \int_\Omega p_1(x,y)dx\, dy, \qquad \text{(either } q_2 \text{ and } q\text{)},$$

which are also approximated by a numerical calculation of these integrals. We used for our example

$$p_1(x,y) = 1 + \frac{(x+y)^2}{10}, \quad q_1 = 2.13,$$

$$p_2(x,y) = 1 + \frac{1}{(1+x+y)^2}, \quad q_2 = 1.09,$$

$$p(x,y) = 0, \quad q = 0, \quad \Omega = (0,1)x(0,1).$$

It should be noted that such a preconditioning guarantees the fact (5.1.5) with constant β_1/β_n but for more oscillating coefficients p_1 and p_2 the condition of $C_o^{-1}K$ could probably be a large number. Another class of preconditionings are the incomplete Cholesky factorizations IC(0) (compare [SCHWARZ 4, AXELSSON 1]), where

$$K = \tilde{L}\,\tilde{L}^T + N,$$

$$C_2 = \tilde{L}\,\tilde{L}^T$$

and $\tilde{L}$ has the same sparsity structure as K. Due to [GUSTAFSSON 1] the fact

$$\varkappa(C_2^{-1}K) = O(h^{-2})$$

arises. A third preconditioning C_1 with

$$\varkappa(C_1^{-1}K) = O(h^{-1})$$

is constructed with the help of the modified incomplete Cholesky factorization of K [GUSTAFSSON 1, 2, 3]. For 5-point-difference-matrices K this method MIC(0) coincides with the method of approximate factorization (MAF) due to [KUČEROV/MAKAROV 1], that is, if

$$K = \text{diag}\,(K) + L + L^T,$$

then

$$C_1 = (D+L)D^{-1}(D+L^T)$$

and the diagonal matrix D is defined from the identity

$$C_1 e = Ke, \quad e = (1,\ldots,1)^T.$$

The numerical results obtained with PSGM using

a) no preconditioning $C = I$,

b) $C = C_o$, here $\varkappa(C^{-1}K) = O(1)$,

c) $C = C_1$, here $\varkappa(C^{-1}K) = O(h^{-1})$,

d) $C = C_2$, here $\varkappa(C^{-1}K) = O(h^{-2})$,

and

e) $C = K$ $(\varkappa(\Theta) = 1)$

are presented in Table 1.

(In each case we calculated 4 eigenvalues using $p = 5$.) For a better understanding of the results in the table we stress the main difference between a) ... d) and e). The computational costs of one step of PSGM (K,M,C) are a multiple of n for the cases a) to d) (a solution with C_0 using FFT would need $O(n \cdot \ln n)$ operations). For the case e) we calculated the Cholesky-decomposition of K before starting the iteration process. For finer discretization this requires a back-storage and represents a very time consuming procedure. The total complexity of the decomposition itself becomes $O(h^{-4})$ and one iteration step of the gradient method needs then $O(h^{-3})$ for the forward- and backward substitution. Hence, at least the case b) must have smallest computational costs (for h small enough), which are growing with $O(h^{-2})$ only.

Also the preconditioning C_1 can be recommended. Here, the total number of iterations is slightly growing with $h \rightarrow 0$, but the preconditioning itself is considerably cheap.

Note that $C = K$ is the "best choice" of C from the standpoint of the convergence quotients η_j, because $C = K$ yields the smallest quotients for the given problem $\{K,M\}$, namely

$$\eta_j = \left(\frac{1 - \gamma_j}{1 + \gamma_j}\right)^2, \qquad j = 1,\dots,p.$$

On the other hand, there arises the numerically difficult problem to solve linear systems with K at any step of the iteration. (This should also be done using preconditionings for larger problems.)

Nevertheless, the choice $C = K$ enables further simplifications of the iteration process, thus it will be discussed in the next chapter. The results for $C = K$ shown in Table 1 are obtained from this more effective realization ($SGM(K^{-1}M)$).

Algorithm \ Dimension n		15 x 15	31 x 31	63 x 63
Cholesky-decomposition of K		1" (8R) (8W)	16" (261R) (67W)	4'10" (5544R) (559W)
Simultaneous Iteration		**19** 27" (304R)	**25** 4'19" (3350R)	**23** 29'50" (25714R)
Simultaneous Gradient Method	C = K	**10** 19" (160R)	**10** 1'55" (1340R)	**10** 13'36" (11180R)
	C = I	**65** 1'55"	>200 >22	-
	$C = C_2$	**14** 29"	**32** 4'16"	
	$C = C_1$	**11** 23"	**20** 2'43"	**29** 16'08"
	$C = C_0$	**13** 37"	**16** 2'59"	**15** 11'42"

Table 1

The comparison of the results for various preconditionings in PSGM(K,M,C)

(Note that the time required for the Cholesky-decomposition of K is to be added in the rows "Simultaneous Iteration" and "C = K" for obtaining full information. Here, R (resp. W) means read-(write-)operations in using back storage, when the Cholesky-decomposition is performed for a variable bandwidth storage scheme of K divided into records up to 3600 Bytes.)

5.2. The Simultaneous Gradient Method for the Special Eigenvalue Problem of $\mathcal{A} = K^{-1}M$

According to Chapter 5.1., the choice C = K leads to an algorithm for the calculation of the p largest eigenvalues and corresponding eigenvectors of the operator $\mathcal{A} = K^{-1}M$. Also the choice C = M would yield a special eigenvalue problem. Here, the p largest eigenvalues of the problem

$$\mathcal{A} = \mathcal{J}, \quad \mathcal{B} = M^{-1}K$$

coincide with the p smallest of the operator $\tilde{\mathcal{A}} = M^{-1}K$ (if both K and M are positive definite). On the other hand, the latter choice will lead to an arbitrary slow convergence, because η_i tend to unity $(i=1,\dots,p)$ for finer discretizations. If the problem

$$K\varphi = \omega^2 M\varphi$$

has eigenvalues

$$\omega_1^2 \leq \dots \leq \omega_n^2,$$

the "gap-sizes" $\tilde{\gamma}_i$ for the operator $\tilde{\mathcal{A}} = M^{-1}K$ become

$$\tilde{\gamma}_i = \frac{\omega_{p+1}^2 - \omega_i^2}{\omega_n^2 - \omega_i^2} \longrightarrow 0 \qquad \text{(for } h \longrightarrow 0\text{)}.$$

The eigenvalues of $\mathcal{A} = K^{-1}M$ are the inverses of ω_i^2. This causes the gaps

$$\gamma_i = \frac{\frac{1}{\omega_i^2} - \frac{1}{\omega_{p+1}^2}}{\frac{1}{\omega_i^2} - \frac{1}{\omega_n^2}} = \frac{\omega_n^2}{\omega_i^2} \cdot \tilde{\gamma}_i,$$

which are enlarged about a factor near $\varkappa(\mathcal{A})$ compared with $\tilde{\gamma}_i$. Hence, the choice C = M is unsatisfactory and will not be discussed anymore. This fact agrees with the observations in [SCOTT 1].

The simultaneous gradient method for maximizing $\mu(X)$ in order to calculate p largest eigenvalues of the operator $\mathcal{A}$ has been described in Chapter 3.4. From this chapter the problem of calculating the two matrices

$$E_2 = X^H \mathcal{A}^2 X \qquad \text{and} \qquad E_3 = X^H \mathcal{A}^3 X$$

for defining the Ritz-step was left to solve. The powers of the operator

$\mathcal{A}$ in $E_{2/3}$ do not involve any difficulties, if a proper inner product is defined as follows.
Let K be positive definite. For any function $f(\tau)$, defined on the eigenvalues of $\mathcal{A}$ with

$$f(\lambda_i) > 0, \qquad i = 1,\dots,n,$$

the matrix

$$\mathcal{T} = Kf(K^{-1}M) \tag{5.2.1}$$

is symmetric, positive definite and $\mathcal{A}$ becomes a symmetric operator in the $\mathcal{T}$-inner product (compare [MEYER 1]):

$$\begin{aligned}\langle \mathcal{A}x,y\rangle &= y^T\mathcal{T}\mathcal{A}x = y^TK\, f(K^{-1}M)K^{-1}Mx\\ &= y^TKK^{-1}M\, f(K^{-1}M)x\\ &= (K^{-1}My)^TKf(K^{-1}M)x\\ &= \langle x,\mathcal{A}y\rangle .\end{aligned}$$

For positive definite M and possibly indefinite K the matrix

$$\mathcal{T} = M\,\tilde{f}(M^{-1}K)$$

can be used for a proper definition, if $\tilde{f}(\tau)$ is positive over the eigenvalues of $M^{-1}K$. If both K and M are positive definite, then both definitions coincide and, obviously,

$$f(\tau) = \tau\tilde{f}(\tau^{-1}).$$

Now, the function $f(\tau)$ will be used to simplify the expressions of E_2 and E_3. A convenient choice, if M is positive definite, consists in

$$\tilde{f}(\tau) = \tau^2 \qquad \text{resp. } f(\tau) = \tau^{-1}$$

which yields

$$\mathcal{T} = KM^{-1}K.$$

Then we are able to determine E_2 and E_3 during the main step, the calculation of

$$Z = \mathcal{A}X.$$

The definitions

$$Y = MX$$

and

$$Z = K^{-1}Y$$

(that is, Z is the solution of KZ = Y) lead to

$$E_2 = X^H \mathcal{A}^2 X = X^T \mathcal{T} \mathcal{A}^2 X = X^T K \mathcal{A} X = \underline{\underline{X^T Y}},$$

$$E_3 = X^H \mathcal{A}^3 X = X^T \mathcal{T} \mathcal{A}^3 X = X^T K \mathcal{A}^2 X = X^T M K^{-1} M X = \underline{\underline{Y^T Z}}.$$

Hence, for positive definite matrices M the simultaneous gradient method for the calculation of p largest eigenvalues of the operator $\mathcal{A} = K^{-1}M$, using $KM^{-1}K$-inner product, reads as follows:

<u>Algorithm $SGM_1(K^{-1}M)$</u> (for $M > 0$)

(1) $Y := MX$,

$E_2 := X^T Y$.

(2) Solve for Z:

$$KZ = Y,$$

$E_3 := Y^T Z$.

(3)

$$A = \begin{pmatrix} \Delta & E_2 \\ E_2 & E_3 \end{pmatrix}, \qquad B = \begin{pmatrix} I & \Delta \\ \Delta & E_2 \end{pmatrix},$$

calculate p largest eigenvalues
$\hat{\delta}_1 \geq \ldots \geq \hat{\delta}_p$ with corresponding eigenvectors
$q_1 \ldots q_p$ of $\{A, B\}$,

$$AQ = BQ\hat{\Delta}, \quad Q^T BQ = I.$$

(4) $\hat{X} := (X \vdots Z)Q$.

<u>Remarks</u>

1. Note that E_2 and E_3 are symmetric matrices. Hence, only p(p+1)n multiplications are required for forming the Ritz-step.
2. For the implementation of that algorithm Chapter 6 should be considered to avoid numerical instabilities near the solution Φ_1.
3. For indefinite M the function $f(\tau) = \tau^{-1}$ is not allowed, because some negative eigenvalues of $\mathcal{A}$ exist (provided that K is positive definite for that case). Here, we successfully used $f(\tau) = \tau^{-2}$.

For $f(\tau) = \tau^{-2}$, thus $\mathcal{F} = K\mathcal{A}^{-2}$, the direct calculation yields

$$E_3 = X^T\mathcal{F}\mathcal{A}^3X = X^TK\mathcal{A}X = X^TY,$$

while E_2 is to be updated from $\check{E}_2$, $\check{E}_3$ and $\check{E}_4$.
With

$$E_4 = X^H\mathcal{A}^4X = X^TK\mathcal{A}^2X = Y^TZ$$

the matrix $\hat{E}_2$ for the next step of the iteration is determined using

$$\hat{X} = (X \vdots \mathcal{A}X)Q$$

as

$$\hat{E}_2 = \hat{X}^H\mathcal{A}^2\hat{X} = Q^T(X \vdots \mathcal{A}X)^H\mathcal{A}^2(X \vdots \mathcal{A}X)Q$$
$$= Q^T\begin{pmatrix} E_2 & E_3 \\ E_3 & E_4 \end{pmatrix} Q.$$

Hence, $\hat{E}_2$ will be calculated after step (3).

Algorithm $SGM_2(K^{-1}M)$ (for $K > 0$, M arbitrary)

(1) $Y := MX$,

$E_3 := X^T Y$.

(2) Solve for Z

$$KZ = Y,$$

$E_4 := Y^T Z$.

(3)
$$A = \begin{pmatrix} \Delta & E_2 \\ E_2 & E_3 \end{pmatrix}, \qquad B = \begin{pmatrix} I & \Delta \\ \Delta & E_2 \end{pmatrix},$$

calculate p largest eigenvalues

$\hat{\delta}_1 \geq \ldots \geq \hat{\delta}_p$ with corresponding eigenvectors

$q_1 \ldots q_p$ of $\{A, B\}$,

$$AQ = BQ\hat{\Delta}, \quad Q^T BQ = I.$$

(4) $$\hat{E}_2 := Q^T \begin{pmatrix} E_2 & E_3 \\ E_3 & E_4 \end{pmatrix} Q.$$

(5) $\hat{X} := (X \vdots Z)Q$.

Remark to step (2) of both algorithms

If the linear systems $KZ = Y$ are solved by using the Cholesky-decomposition of K we have a more ingenious possibility for the calculation of $Y^T Z$.

Let L be the lower Cholesky-factor of K,

$$K = LL^T,$$

then $KZ = Y$ is solved in two steps

1) $L\tilde{Z} = Y$,

2) $L^T Z = \tilde{Z}$,

which causes the following identity: $Y^T Z = Z^T K Z = (L^T Z)^T (L^T Z) = \tilde{Z}^T \tilde{Z}$.

Hence, the matrix $Y^T Z = \tilde{Z}^T \tilde{Z}$ is obtained between the forward- and backward-substitution. This fact allows that $\tilde{Z}$ will be written over Y and Z over $\tilde{Z}$ and the whole algorithm requires the storage for two rectangular matrices only.

5.3. Spectral Transformation

All the simultaneous algorithms described here will show at least two disadvantages.

At first, if the dimension of the iterated subspaces span X is chosen to be p, then the total computational costs are a multiple of p^2n (from forming the inner product matrices and from the last step $\hat{X} = (X\vdots S)Q$). The storage requirements are linearly growing with p. Hence, for the calculation of a larger number of eigenvalues these algorithms become more problematic.

Secondly, the rate of convergence depends on the eigenvalue distribution and is unknown for each new problem. A very rapid rate of convergence can be produced by a change of the gap-sizes γ_i only, that is, by transforming the eigenvalues of the given problem. Unfortunately, the spectral transformation in connection with the preconditioning ideas of Chapter 5.1. has appeared to be impossible up to now.

These new algorithms will overcome the disadvantages described above, but they will require solutions of indefinite linear systems of equations, which represents a new difficult numerical problem. The main idea for using spectral transformation is the application of any algorithm which computes largest eigenvalues of an operator $\mathcal{A}$, when $\mathcal{A}$ is replaced by an new operator $\mathcal{C}$, a function of $\mathcal{A}$.

From 5.2. we may recall the given problem

$$K\varphi = \omega^2 M\varphi$$

with the eigenvalues

$$\omega_1^2 < \omega_2^2 < \ldots < \omega_n^2 ,$$

respectively

$$\mathcal{A}\varphi = \lambda\varphi \qquad (\mathcal{A} = K^{-1}M)$$

with the eigenvalues

$$\lambda_i = \frac{1}{\omega_i^2} .$$

At least two functions of $\mathcal{A}$ can be successfully used for an accelerating spectral transformation, namely

$$f_1(\tau) = \frac{\tau}{1 - \sigma\tau} \tag{5.3.1}$$

and

$$f_2(\tau) = \frac{1}{1 - \sigma\tau} . \tag{5.3.2}$$

Here, we obtain

$$\mathcal{C}_1 = f_1(\mathcal{A}) = (I-\sigma\mathcal{A})^{-1}\mathcal{A}$$

$$= (I-\sigma K^{-1}M)K^{-1}M \qquad (5.3.3)$$

$$= (K-\sigma M)^{-1}M,$$

$$\mathcal{C}_2 = f_2(\mathcal{A}) = (I-\sigma\mathcal{A})^{-1}$$

$$= (K-\sigma M)^{-1}K. \qquad (5.3.4)$$

Obviously, σ should not agree with one of the eigenvalues ω_i^2. The eigenvalues of $\mathcal{C}_1$ (resp. $\mathcal{C}_2$) are

$$f_1(\lambda_i) = \frac{1}{\omega_i^2 - \sigma}$$

$$\text{resp. } f_2(\lambda_i) = \frac{\omega_i^2}{\omega_i^2 - \sigma} .$$

Hence, the eigenvalues ω_i^2 near σ are transformed into absolutely very large eigenvalues of $\mathcal{C}_{1/2}$.
Furthermore,

$$\mathcal{C}_2 = (K-\sigma M)^{-1}(K-\sigma M+\sigma M)$$

$$= \mathcal{I} + \sigma\mathcal{C}_1,$$

thus the new gap-sizes of the transformed eigenvalues are identical for both $\mathcal{C}_1$ and $\mathcal{C}_2$ (they are independent of shifts and scaling). For the same reason, the convergence of the simultaneous method of steepest ascent (of its accelerations and of the Lanczos-algorithm) is identical for both operators. The simple Simultaneous Iteration (see 3.5) would have a slightly different behaviour with respect to $\mathcal{C}_1$ or $\mathcal{C}_2$, but for some other reason this method will not be recommended.
This spectral transformation leads to the possibility of calculating a larger number of eigenvalues, when a series of shifts σ is used. For each single shift σ we calculate a small number of eigenvalues of $\mathcal{C}$ ($\mathcal{C}$ stands for either $\mathcal{C}_1$ or $\mathcal{C}_2$), which are very large in absolute magnitude and thus very quickly convergent. Both operators $\mathcal{C}_1$ and $\mathcal{C}_2$ require the solution of the (possibly) indefinite systems

$$(K-\sigma M)Z = Y,$$

hence, for each shift σ the indefinite factorization

$$(K - \sigma M) = LDL^T$$

could be calculated before starting the eigenvalue iteration process. Here, D denotes an indefinite diagonal matrix (or block-diagonal matrix for increasing the stability, compare [DUFF, et al. 1]), which yields an important information about the number of eigenvalues ω_i^2 smaller than σ. From the Sylvester law of inertia, the number n_σ of negative entries in D is equal to the number of eigenvalues ω_i^2 fulfilling

$$\omega_i^2 \in (-\infty, \sigma) \qquad \text{for } M > 0,\ \sigma \text{ arbitrary,}$$

resp. $\omega_i^2 \in (0, \sigma)$ for indefinite M and $\sigma > 0$.

On the other hand, we have the number m_σ of eigenvalues, which have been calculated up to now. Hence, for the new fixed shift σ we need

$$n_\sigma - m_\sigma = q_\sigma$$

new eigenvalues, which belong to negative, especially to the q_σ smallest eigenvalues of $\mathfrak{C}$. If we again denote the size of the iterated subspaces by p, then in addition to q_σ, we will be able to calculate $p_\sigma = p - q_\sigma$ [1)] positive, largest eigenvalues of $\mathfrak{C}$, which correspond to next ω_i^2 larger than σ.
Thus, for the actual shift σ we will calculate q_σ smallest and p_σ largest eigenvalues of $\mathfrak{C}$ with corresponding eigenvectors. This agrees with the maximization of the functional $\mu(X_1, X_2)$, $X_1 \in \mathbf{H}^{p_\sigma}$, $X_2 \in \mathbf{H}^{q_\sigma}$, according to Chapter 2.4. Here,

$$F(X) = (X^H X)^{-1} X^H \mathfrak{C} X$$

and

$$\mu(X_1, X_2) = \mathrm{tr}F(X_1) - \mathrm{tr}F(X_2).$$

If the simple Simultaneous Iteration was used, we would never have this easy control to select q_σ negative and p_σ positive eigenvalues of $\mathfrak{C}$, because this method yields p absolutely largest, which would not necessarily be these eigenvalues expected. This control is offered by all algorithms based on Ritz-steps. [ERICSSON/RUHE 1] described spectral

---

1) Needless to say that the fact $p - q_\sigma < 0$ requires a change of σ. With a new (smaller) shift the process is repeated.

transformations in connection with the Lanczos-algorithm, however some authors stressed that for the very large eigenvalue gaps, as they arise here, the complicate Lanczos-algorithm does not seem to be necessary and a simple power method (Simultaneous Iteration for some more eigenvalues) was sufficient. Our simultaneous method of steepest ascent for $\mu(X_1,X_2)$ is lying between Simultaneous Iteration and Lanczos, but it will be recommended because of the Ritz-step for selecting the expected eigenvalues.

For the construction of the simultaneous method of steepest ascent for the functional $\mu(X_1,X_2)$ we recall from 2.4. the gradient

$$\nabla \mu(X_1,X_2) = (2R(X_1) \vdots -2R(X_2)).$$

Here, $X = (X_1 \vdots X_2) \in \mathbb{H}^p$ is again assumed to be a Ritz-basis of the subspace span X, meaning

$$X^H X = I \quad \text{and} \quad X^H \mathcal{C} X = \Delta.$$

According to 3.1. the direction of movement could be chosen as

$$S = \nabla \mu(X_1,X_2).$$

But again, the result of the Ritz-step using X and S consists in

$$\hat{X} = (\hat{X}_1 \vdots \hat{X}_2) = (X \vdots S)Q$$

and $\hat{\Delta}$ (eigenvalue approximations), which are independent of the special basis $(X \vdots S)$ of that subspace. Hence, due to

$$\begin{aligned} \text{span}\,(X \vdots S) &= \text{span}\,(X \vdots R(X_1) \vdots + R(X_2)) \\ &= \text{span}\,(X \vdots R(X)) \\ &= \text{span}\,(X \vdots \mathcal{C} X) \end{aligned}$$

we will calculate the Ritz-projection for the latter basis $(X \vdots \mathcal{C}X)$ and the algorithm agrees with the simultaneous gradient method of Chapter 3.4. except for the selection of the Ritz-eigenvalues of $\{A,B\}$ and the corresponding eigenvectors in $Q \in \mathbb{R}^{2p\times p}$. Here, for maximizing $\mu(X_1,X_2)$ we will select q_c smallest and p_c largest eigenvalues of $\{A,B\}$, which are the new eigenvalue approximations in $\hat{\Delta}$ (compare Lemma 2.4.4). Hence, the following algorithm arises:

<u>Simultaneous Gradient Method for Maximizing</u> $\mu(X_1,X_2)$ <u>for the Operator</u> $\mathcal{C}$,

$(X_1 \in \mathbb{R}^p,\ X_2 \in \mathbb{R}^q,\ p = p_{\mathcal{C}} + q_{\mathcal{C}})$

(1) $Z := \mathcal{C}X$.

(2) $E_2 := X^H \mathcal{C}^2 X$,

$E_3 := X^H \mathcal{C}^3 X$.

(3) $A = \begin{pmatrix} \Delta & E_2 \\ E_2 & E_3 \end{pmatrix}$, $\qquad B = \begin{pmatrix} I & \Delta \\ \Delta & E_2 \end{pmatrix}$.

Solve the eigenvalue problem of $\{A,B\}$, eigenvalues: $\delta'_1 \leq \ldots \leq \delta'_{2p}$,
choose $\hat{\delta}_1 = \delta'_1, \ldots, \hat{\delta}_{q_{\mathcal{C}}} = \delta'_{q_{\mathcal{C}}}$ ($q_{\mathcal{C}}$ smallest eigenvalues)
and

$$\hat{\delta}_{q_{\mathcal{C}}+1} = \delta'_{2p}, \quad \hat{\delta}_{q_{\mathcal{C}}+2} = \delta'_{2p-1}, \ldots,$$

$$\hat{\delta}_p = \delta'_{2p-p_{\mathcal{C}}+1} = \delta'_{p+q_{\mathcal{C}}+1} \quad (p_{\mathcal{C}} \text{ largest eigenvalues}).$$

Let $Q \in \mathbb{R}^{2p \times p}$ contain the eigenvectors belonging to the selected eigenvalues $\hat{\Delta}$.

$$AQ = BQ\hat{\Delta}, \quad Q^T BQ = I.$$

(4) $\hat{X} = (X \vdots Z)Q$.

In order to apply this basic algorithmic idea to the operators $\mathcal{C}_1$ and $\mathcal{C}_2$ we have to define a proper inner product again, especially for an easy calculation of E_2 and E_3. The operator $\mathcal{C}$ was constructed as a function of $\mathcal{A}$, hence $\mathcal{C}$ will be symmetric, if $\mathcal{A}$ is. That is, we will use the inner product

$$\langle x,y \rangle = y^T \mathcal{F} x$$

with

$$\mathcal{F} = Kf(\mathcal{A}) = M\tilde{f}(\mathcal{A}^{-1})$$

according to Chapter 5.2.
For this definition a difference between $\mathcal{C}_1$ and $\mathcal{C}_2$ occurs. Let us begin with $\mathcal{C}_1$. Here, we assume M to be positive definite.
Note that the first step of our algorithm for $\mathcal{C} = \mathcal{C}_1$ becomes:

(1) $Y := MX$.

Solve for Z: $(K - \sigma M)Z = Y$.

Then we define

$$\tilde{f}(\tau) = \tau^2(1-\frac{\sigma}{\tau})^2 > 0. \tag{5.3.5}$$

This yields

$$\mathcal{F} = M\mathcal{A}^{-2}(\mathcal{I}-\sigma\mathcal{A})^2 = M\mathcal{C}_1^{-2} \tag{5.3.6}$$

$$(= (K-\sigma M)M^{-1}(K-\sigma M))$$ [1]

and

$$E_2 = X^H \mathcal{C}^2 X = X^T\mathcal{F}\mathcal{C}^2X = X^TMX = \underline{\underline{X^TY}},$$

$$E_3 = X^H\mathcal{C}^3X = X^TM\mathcal{C}X = Y^T(K-\sigma M)^{-1}Y = \underline{\underline{Y^TZ}}.$$

For indefinite M the definition (5.3.6) would lead to an indefinite matrix $\mathcal{F}$, which is not allowed. We will consider the operator $\mathcal{C}_2$ for that case. Then both operators $\mathcal{C}_1$ and $\mathcal{C}_2$ are a product of a symmetric matrix $((K-\sigma M)^{-1})$ with a positive definite one (M or K resp.). For $\mathcal{C}=\mathcal{C}_2$ the first step of the above algorithm calculates:

(1) Y:= KX.

Solve for Z: $(K-\sigma M)Z = Y$.

Here, the definition

$$f(\tau) = (1-\sigma\tau)^2 \tag{5.3.7}$$

leads to the same formulas as for $\mathcal{C}=\mathcal{C}_1$:
We obtain

$$\mathcal{F} = K(\mathcal{I}-\sigma\mathcal{A})^2 = K\mathcal{C}_2^{-2} \tag{5.3.8}$$

$$(= (K-\sigma M)K^{-1}(K-\sigma M))$$

and

$$E_2 = X^H\mathcal{C}_2^2X = X^TKX = \underline{\underline{X^TY}},$$

$$E_3 = X^H\mathcal{C}_2^3X = X^TK\mathcal{C}X = Y^T(K-\sigma M)^{-1}Y = \underline{\underline{Y^TZ}}.$$

---

1) We used $\mathcal{A} = K^{-1}M$ for purposes of brevity in this calculations, but for $\mathcal{C}_1$ as well as for the definition of $\mathcal{F}$ the matrix K would be allowed to be singular.

The difference in the application of either $\mathcal{C}_1$ or $\mathcal{C}_2$ consists in the first algorithmic step only. Either Y is calculated as MX or as KX. This leads to a further generalization for definite matrix pencils $\{K,M\}$.
In this case neither K nor M is positive definite, but there exists a positive definite linear combination of K and M, say $\alpha K + \beta M > 0$.
Then

$$\mathcal{C}_3 = (K - \sigma M)^{-1} (\alpha K + \beta M) = \alpha \mathcal{C}_2 + \beta \mathcal{C}_1$$

has the eigenvalues

$$\frac{\alpha \omega_i^2 + \beta}{\omega_i^2 - \sigma}$$

and is a symmetric operator in the $\mathcal{T}$-inner product for

$$\mathcal{T} = (\alpha K + \beta M) \mathcal{C}_3^{-2}$$

$$= (K - \sigma M) (\alpha K + \beta M)^{-1} (K - \sigma M).$$

This again yields

$$E_2 = X^H \mathcal{C}_3^2 X = X^T(\alpha K + \beta M) X = \underline{\underline{X^T Y}},$$

$$E_3 = X^H \mathcal{C}_3^3 X = X^T(\alpha K + \beta M) \mathcal{C}_3 X = \underline{\underline{Y^T Z}},$$

if the element Y is calculated according to

$$Y = \alpha MX + \beta KX.$$

Hence, the operators $\mathcal{C}_1$ and $\mathcal{C}_2$ (and thus the corresponding algorithms) are contained as a special case, for:

$$\alpha = 0, \quad \beta = 1 \implies \mathcal{C}_1,$$

$$\alpha = 1, \quad \beta = 0 \implies \mathcal{C}_2.$$

The final algorithm has the following form:

Simultaneous Gradient Method Using Spectral Transformation ($\alpha K + \beta M > 0$)

(1) $Y := \alpha KX + \beta MX,$

$E_2 := Y^T X.$

(2) Solve for Z:

$(K - \sigma M)Z = Y,$

$E_3 := Y^T Z.$

(3) $A = \begin{pmatrix} \Delta & E_2 \\ E_2 & E_3 \end{pmatrix}, \qquad B = \begin{pmatrix} I & \Delta \\ \Delta & E_2 \end{pmatrix}.$

Solve the eigenvalue problem of $\{A,B\}$, select q_σ smallest and p_σ largest eigenvalues into $\hat{\Delta}$, the corresponding eigenvectors into Q,

$AQ = BQ\hat{\Delta}, \quad Q^T BQ = I.$

(4) $\hat{X} = (X \vdots Z)Q.$

Concluding Remarks

1. We have considered the calculation of p extremal eigenvalues of $\mathcal{T}$ using one shift σ here. For determining a large number of eigenvalues ω_1^2 of the original problem $\{K,M\}$ a strategy for choosing the shifts is required. This problem has been investigated by [ERICSSON/RUHE 1] or [LEZIUS 1]. We recommend the calculation of less than p eigenvalues for one shift (up to the expected accuracy). The remaining eigenvalue approximations (with low accuracy) are used for a new guess of the next σ. More precisely, we will certainly calculate the q_σ smallest (negative) eigenvalues of $\mathcal{T}$. The remaining $p_\sigma = p - q_\sigma$ positive eigenvalues are divided into two groups: Some (the largest, which are rapidly convergent) will be iterated up to the expected accuracy. The other eigenvalues will give us sufficient information about the next part of the spectrum and determine the new σ (after transforming back to approximations for ω_1^2) We found that we should never set the new σ to a very accurate eigenvalue approximation, we should better choose σ between two approximated eigenvalues. Note that the approximate accuracy of all eigenvalue approximations is known from their relative change in the iteration process (as will be seen more clearly in Chapter 6.2).

2. This algorithm causes some problems in avoiding instabilities. The reason is the possibly very large variation of the absolute size of the eigenvalues of $\mathcal{C}$. In all other algorithms the first p eigenvalues of $\mathcal{A}$ or $\{\mathcal{A},\mathcal{B}\}$ vary by a factor 10 ... 100 only, which is uncritical. Here, the spectral transformation could lead to factors up to 10^5 or more, if σ was chosen too near to an ω_1^2.
Hence, for the implementation of that algorithm the comments in Chapter 6 must be taken into consideration.

6. SOME ASPECTS OF THE IMPLEMENTATION OF THE PREVIOUSLY DESCRIBED ALGORITHMS

A stable calculation of the eigenvalues and eigenvectors of the small pencils $\{A,B\}$ arising from the Ritz-steps of our algorithms represents an important problem of the implementation. This will be discussed in the first part of this chapter. We are going to demonstrate a solution technique, which also detects the dangerous (near-)linear dependence of the columns defining A and B.
The second part is devoted to the stopping criterion of the algorithms. The last part of this chapter contains an important comment on the simplest algorithm, the gradient method, which can be substantially accelerated by the use of overrelaxation. We create an automatic choice of overrelaxation factors for increasing the projection onto the desired eigenspace.

6.1. The Solution of $\{A,B\}$

Due to the fact that the matrices A and B, arising from the Ritz-projection technique at each step of the iteration process, are small and dense, the solution of the eigenvalue problem

$$AQ = BQ\hat{\Delta}$$

does not seem to cause any complications. There are some well-known methods (compare [EISPACK 1 and 2]) for this problem. The simplest one is the transformation of $\{A,B\}$ into an equivalent special eigenvalue problem of the matrix $\tilde{A}$ using the Cholesky-decomposition of B. (Note that B is always positive definite for our algorithms.)
Let L be the lower Cholesky-factor of B,

$$B = LL^T \tag{6.1.1}$$

then

$$Aq = \hat{\delta} Bq$$

is equivalent to

$$\tilde{A}(L^Tq) = \hat{\delta}\,(L^Tq)$$

with

$$\tilde{A} = L^{-1}AL^{-T}. \tag{6.1.2}$$

The implementation of this method requires the two matrices A and B in the readily accessible storage only. L is written over B, $\tilde{A}$ is written over A. Furthermore, the eigenvalues and eigenvectors of $\tilde{A}$ are calculated by Householder-tridiagonalization of $\tilde{A}$, using the implicit QL-algorithm.

Here, the eigenvector matrix Q is written over $\tilde{A}$, again.
If the problem $\{A,B\}$ has absolutely small and large eigenvalues together, some criticism of that method will be provoked (compare [STEWART 1]). In this case the smaller one may be rather disturbed, whenever the condition of B is larger. For the algorithms, here, this method is recommended, because only the p largest eigenvalues are of interest. (These should be the absolutely largest for later steps of the iteration, as well. For the gradient method with spectral transformation of 5.3. the p eigenvalues are ideed, approximately the absolutely largest.) In using this method the calculation of the large eigenvalues of $\{A,B\}$ will be stable, if the condition of B is not too large.
Another advantage derives here from the fact that A and B are Ritz-projections. That is, for an element $Y \in \mathbf{H}^{p+m}$ we have

$$A = Y^H \mathcal{A}\, Y, \qquad B = Y^H \mathcal{B}\, Y$$

for certain symmetric operators and a special inner product. In any case, Y is divided into two parts. The first p columns of Y are the columns of the iterated element $X \in \mathbf{H}^p$.
The remaining columns are:

a) the residual $R = \mathcal{A}X - \mathcal{B}X\Delta$
   for the simultaneous method of steepest ascent ($\mathcal{B} \neq \mathcal{I}$), e.g. PSGM(K,M,C),
b) the element $Z = \mathcal{A}X$
   for the simultaneous method of steepest ascent for $\mathcal{B} = \mathcal{I}$, e.g. $SGM_{1/2}(K^{-1}M)$ or $SGM(\mathcal{C})$,
c) linear combinations of R (resp. Z) and W for the simultaneous method of conjugate gradients (see 3.2),
d) R (resp. Z) together with W for the simultaneous W-accelerated method of steepest ascent (see 3.3).

As a general problem of all these cases, the columns of Y become more and more linearly dependent during the iteration process. Especially, the cases b) and d) (for $\mathcal{B} = \mathcal{I}$) are dangerous, because from $X \rightarrow \Phi_1$ follows $\mathcal{A}X \rightarrow \Phi_1 \Lambda_1$ and Y has some more and more parallel columns. Hence, A and B would be near to a singular pencil with common null-space, if not some crucial columns of Y are dropped. The near-singular-pencil-property of $\{A,B\}$ would destroy the ordered convergence of the eigenvalue approximations Δ. Especially, one or two useless (spurious) eigenvalues in $\hat{\Delta}$ would occur at some steps.
For the case a) the columns of R tend to zero, which is also a kind of near-linear dependence of the columns of Y. This is less dangerous than b), because small columns in Y could be enlarged by some factors. This is equivalent to a diagonal scaling of $\{A,B\}$ to

$$\{DAD,\ DBD\} \quad (\text{i.e. } DBD = (YD)^H \mathcal{B}\,(YD))$$

and could lead to more equilibrated pencils. The residual R is calculated by cancelations, hence, some columns of R represent random-vectors for later steps of the iteration. On the other hand, if the inner products $Y^H \mathcal{A} Y$ and $Y^H \mathcal{B} Y$ are calculated without using some orthogonalities (such as $X^H R = \mathcal{O}$ for $\mathcal{B} = \mathcal{I}$) then $\{A, B\}$ represents a Ritz-projection into span Y (numerically good enough). For both situations a) and b) the same tool according to the next considerations will be successfully used.

In order to describe a technique for avoiding the resulting instabilities we will add one important fact here:

The Gram-Schmidt-orthogonalization ($\mathcal{B}$-inner product) of the columns of Y would detect such near-linear dependence. Let $\tilde{Y}$ be the result of the Gram-Schmidt-orthonormalization process,

$$\tilde{Y} = (\tilde{y}_1, \ldots, \tilde{y}_p), \tag{6.1.3}$$

$$\tilde{Y}^H \mathcal{B} \tilde{Y} = I, \tag{6.1.4}$$

$$\begin{aligned} l_{jj}\tilde{y}_j &= y_j - \sum_{i<j} l_{ji}\tilde{y}_i, \\ l_{ji} &= \langle \mathcal{B} y_j, \tilde{y}_i \rangle, \qquad i < j, \end{aligned} \tag{6.1.5}$$

then

$$\begin{aligned} l_{jj} &= \| y_j - \sum l_{ji}\tilde{y}_i \|_{\mathcal{B}} \\ &= \| (\mathcal{I} - \mathcal{P}_j) y_j \|_{\mathcal{B}} \; . \end{aligned} \tag{6.1.6}$$

Here, $\mathcal{P}_j$ denotes the oblique projection operator onto span $(y_1, \ldots, y_{j-1})$ = span $(\tilde{y}_1, \ldots, \tilde{y}_{j-1})$ orthogonal to span $(\mathcal{B} y_1, \ldots, \mathcal{B} y_{j-1})$ (an orthogonal projection operator with respect to the $\mathcal{B}$-inner product).

$$\mathcal{P}_j = \sum_{i<j} \langle \mathcal{B} \cdot, \tilde{y}_i \rangle \; \tilde{y}_i .$$

Obviously, if

$$l_{jj} \leq \varepsilon_0 \| y_j \|_{\mathcal{B}} \quad \text{for } \varepsilon_0 \text{ small enough,} \tag{6.1.7}$$

then y_j is near-linearly dependent on the columns $y_1, \ldots, y_{j-1}$. But due to the calculation of the Cholesky-factor L of B, the Gram-Schmidt-process is entirely avoided, because the numbers l_{ji} are the entries of L:

$$Y = \tilde{Y}L^T \quad \text{from Gram-Schmidt,}$$

thus

$$B = Y^H B\, Y = L\tilde{Y}^H B\, \tilde{Y}L^T = LL^T.$$

If b_{ij} represent the entries of B, then (6.1.7) is equivalent to

$$l_{jj} \leq \varepsilon_o \cdot \sqrt{b_{jj}}$$

respectively

$$l_{jj}^2 = b_{jj} - \sum_{i<j} l_{ji}^2 \leq \varepsilon_o^2 b_{jj}. \qquad (6.1.8)$$

This inequality can be tested during the Cholesky-decomposition of B for each j.
If (6.1.8) is fulfilled for one j, then the near-linear dependence of y_j on the previous columns is detected. This column will be neglected in further calculations, which need not be done explicitly. It is sufficient that the j-th rows and columns of A and B are dropped and we use smaller submatrices for the transformation into the special eigenvalue problem $\tilde{A}$. In the resulting eigenvectors some zeroes on the j-th position are inserted. Thus, the last step

$$\hat{X} = YQ$$

would multiply y_j by zero. (Q contains p eigenvectors of a subpencil of $\{A,B\}$ filled up with zeroes.) An even simpler technique, which immediately yields this zeroes in Q, consists in the following. The j-th row and column of A is replaced by zeroes, l_{jj} by 1 and all later l_{ij} $(i > j)$ by zero. Then the problem $\{A,B\}$ continues to be solved (without considering special subproblems). Obviously, $\tilde{A}$ gets that zero column and zero row too and has the eigenvalue 0 together with eigenvectors, which are coordinate unit vectors. Hence, all other (our desired) eigenvectors keep this special zero entries immediately. [1)]
(For using this latter technique, we assume that our wanted eigenvalues are non-zeroes, which is common for most problems.)

---

1) If instead of the main diagonal elements l_{jj} their inverses l_{jj}^{-1} are stored, all these replacements directly result from the one definition $l_{jj}^{-1} := 0$, because the j-th row and column of $\tilde{A}$ arises from multiplication by l_{jj}^{-1}, the j-th column of L also.

The problem of choosing the test accuracy ε_o^2 for (6.1.8) is uncritical. Obviously, the larger ε_o has been defined the slower the convergence will arise, because we use smaller subspaces for the Ritz-step. The smaller ε_o has been chosen the more often spurious eigenvalues could occur at the later steps of the iteration. We successfully used

$$\varepsilon_o^2 = 10^{-10} \approx 10^6 \text{ machep}$$

(machep denotes the machine accuracy, here 16^{-13}).
Hence, in the calculation of

$$l_{jj}^2 = b_{jj} - \sum_{i<j} l_{ji}^2$$

a cancelation up to 10 decimal digits is allowed.

Especially for the simultaneous method of steepest ascent for $\mathcal{B} = \mathcal{J}$ (hence, for the operator $\mathcal{A} = K^{-1}M$ or for $\mathcal{C}$ of 5.3) one important device ensures convergence for too large ε_o as well:
If ε_o was chosen too large, then after some steps all columns of $\mathcal{A}X$ from $Y = (X \vdots \mathcal{A}X)$ would be dropped (the resulting dimension of the submatrix problem is decreased to p). Then a further convergence would be impossible. At least for positive definite $\mathcal{A}$ (or for the case of computing absolutely largest eigenvalues as in 5.3) the p-tuple $\mathcal{A}X$ always represents better eigenvector approximations than X (the effect of Simultaneous Iteration). Hence, we better define

$$Y = (\mathcal{A}X \vdots X).$$

This results in a change in the definition of the Ritz-projection matrices A and B. For instance,

$$A_{11} = E_3,\ A_{12} = A_{21} = E_2,\ A_{22} = \Delta,$$

$$B_{11} = E_2,\ B_{12} = B_{21} = \Delta,\ B_{22} = I$$

with $E_i = X^H\mathcal{A}^iX$ (analogously for the operator $\mathcal{C}$ of 5.3). With the help of this altered order of the columns of Y, we would omit the dangerous columns in X, keeping the "better" columns of $\mathcal{A}X$ in the effecting subspace. Hence, even if all columns of X are dropped (for too large ε_o) a further convergence is guaranteed, because at least the Simultaneous Iteration Method will remain (span $\hat{X}$ = span $\mathcal{A}X$). Indeed, for our ε_o-choice the column x_j is dropped near the point, where δ_j is detected to fulfil the expected accuracy.

6.2. Some Comments on the Stopping Criterion

In all our algorithms described above we have not designated the test for convergence. Here, the general point, where this test should be included, and the realization of such a stopping criterion will be discussed.

Some different kinds of convergence tests can be found in the literature. We will deal with the eigenvalue approximations only, which is much cheaper than using residual norms or changes of the eigenvector approximations. In our opinion, it is sufficient, if the following considerations are respected.

Let ε be the user's expected relative accuracy for the eigenvalues (then larger entries of the eigenvectors will have a relative accuracy about $\sqrt{\varepsilon}$). That is, our aim consists in the inequality

$$\alpha_j = \frac{\lambda_j - \delta_j}{\lambda_j} \leq \varepsilon, \qquad j = 1,\dots,p. \tag{6.2.1}$$

Then δ_j will be accepted to be an eigenvalue of $\{\mathcal{A},\mathcal{B}\}$ up to the given accuracy ε.

On the other hand, λ_j is unknown and we are able to consider δ_j and $\hat{\delta}_j$ at one step of the iteration only. Hence, the test of convergence will be made after the solution of the Ritz-problem $\{A,B\}$ in each* of our algorithms, because at this point Δ and $\hat{\Delta}$ are available.

Here, we will calculate the relative change of the eigenvalue approximations

$$\tilde{\alpha}_j = \frac{\hat{\delta}_j - \delta_j}{\hat{\delta}_j}, \qquad j = 1,\dots,p. \tag{6.2.2}$$

If the convergence $\delta_j \rightarrow \lambda_j$ is quick enough, then the test

$$\tilde{\alpha}_j \leq \varepsilon \quad ? \tag{6.2.3}$$

would be sufficient to detect the eigenvalue $\hat{\delta}_j \approx \lambda_j$ to be convergent. For a slower convergence the relation between α_j and $\tilde{\alpha}_j$ must be investigated more deeply. In order to do that we assume an exact linear convergence of δ_j to λ_j (at least asymptotically), that is

$$(\lambda_j - \hat{\delta}_j) = \eta_j(\lambda_j - \delta_j), \qquad j = 1,\dots,p,$$

for later iteration steps. (Here, η_j denotes an unknown but existing

convergence quotient, not its estimation as in Chapter 4.2.)
Obviously,

$$\tilde{\alpha}_j \approx \frac{\hat{\delta}_j - \delta_j}{\lambda_j} = \frac{(\lambda_j - \delta_j) - (\lambda_j - \hat{\delta}_j)}{\lambda_j} = \alpha_j(1 - \eta_j)$$

thus the expected test (6.2.1) could be replaced by

$$\frac{\tilde{\alpha}_j}{1 - \eta_j} \leq \varepsilon \; ? \tag{6.2.4}$$

which enlarges $\tilde{\alpha}_j$ to near α_j, if the unknown quotient η_j could be approximated. The well-known Aitken-technique yields this approximation:

$$\eta_j = \frac{\lambda_j - \hat{\delta}_j}{\lambda_j - \delta_j} = \frac{\lambda_j - \delta_j}{\lambda_j - \check{\delta}_j}$$

thus

$$\frac{\hat{\delta}_j - \delta_j}{\delta_j - \check{\delta}_j} = \frac{(\lambda_j - \delta_j)(1 - \eta_j)}{(\lambda_j - \check{\delta}_j)(1 - \eta_j)} = \eta_j . \tag{6.2.5}$$

This quotient, using the three last eigenvalue approximations $\check{\delta}_j$, δ_j and $\hat{\delta}_j$, can be calculated at each step and if

$$\frac{\hat{\delta}_j - \delta_j}{\delta_j - \check{\delta}_j} < 1 \tag{6.2.6}$$

it will yield an approximation of η_j and can be applied in the final stopping criterion (6.2.4). (Note that (6.2.6) need not be fulfilled for earlier iteration steps.)
Summarizing these facts, the stopping criterion will be the following:
Solution of $AQ = BQ\hat{\Lambda}$

(1) calculate

$$\eta_j = \frac{\hat{\delta}_j - \delta_j}{\delta_j - \check{\delta}_j} , \qquad j = 1,\dots,p,$$

if $\eta_j \geq 1$ (for one j) then (3)

(2) calculate

$$\tilde{\alpha}_j = \frac{\hat{\delta}_j - \delta_j}{\hat{\delta}_j}, \qquad j = 1,\dots,p,$$

if

$$\frac{\tilde{\alpha}_j}{1 - \eta_j} < \varepsilon \qquad \text{for all } j = 1,\dots,p$$

then: finish this algorithmic step and stop
else:

(3) finish this algorithmic step, go to next step.

6.3. Restart and Overrelaxation

All algorithms except the Simultaneous Iteration use the following information about X:

$$X^H \mathcal{B} X = I \qquad \text{and} \qquad X^H \mathcal{A} X = \Delta . \tag{6.3.1}$$

The new iterated element $\hat{X}$ gets the same properties, if the Ritz-eigenvalue problem $\{A,B\}$ represents the Ritz-projection into span $(X \vdots S)$ numerically precise enough. Hence, the facts (6.3.1) will become untrue during the iteration process, if no restarts are introduced, which calculate at least

$$A_{11} = X^H \mathcal{A} X \qquad \text{and} \qquad B_{11} = X^H \mathcal{B} X$$

anew. If the algorithm requires further recurrences (as W-accelerated method of steepest ascent or $SGM_2(K^{-1}M)$), then these should also be refreshed at a restart point.
For the simultaneous methods of steepest ascent (like PSGM (K,M,C) or $SGM_{1/2}(K^{-1}M)$) we recommend to do a restart at each fifth step of the iteration process. For the accelerated versions a restart will consist in a gradient step only, thus we recommend each tenth step here, because otherwise we would considerably destroy the convergence acceleration.
For the method of steepest ascent we are able to combine the restart with an overrelaxation step, which will have a surprising influence on the convergence. This overrelaxation idea is a consequence of numerical tests, where the following remarkable fact was noticed.
From 3.1. we recall

$$R(X) = \mathcal{A} X - \mathcal{B} X \Delta \tag{6.3.2}$$

and

$$\hat{X} = (X \vdots R)Q = XQ_1 + RQ_2 \tag{6.3.3}$$

having maximal value $\mu(\hat{X})$ over span $(X \vdots R)$. From Lemma 2.4.3 we have

$$X^H\hat{R} = R^H\hat{R} = \mathbb{O}$$

for

$$\hat{R} = R(\hat{X}) = \mathcal{A}\hat{X} - \mathcal{B}\hat{X}\hat{\Delta},$$

that is, the columns of two consecutive residuals are orthogonal. Unfortunately, we found that

$$\text{span}\ (\check{X} \vdots \check{R}) \quad \text{and} \quad \text{span}\ (\hat{X} \vdots \hat{R})$$

will not essentially change for later iteration steps and for slow convergent examples. To be more precise, we will illustrate this behaviour for the case p = 1. Here,

$$r = \mathcal{A}x - \delta\mathcal{B}x,$$

$$\delta = \frac{\langle \mathcal{A}x, x\rangle}{\langle \mathcal{B}x, x\rangle} = \mu(x)$$

and without loss of generality

$$\hat{x} = x + \alpha r, \tag{6.3.4}$$

where

$$\mu(\hat{x}) = \max_{\alpha'} \mu(x + \alpha' r).$$

When we calculated the expression

$$\frac{\langle \check{r}, \hat{r}\rangle}{\|\check{r}\|\,\|\hat{r}\|} \tag{6.3.5}$$

at each step of the iteration, we noticed a quick convergence to one. A general proof of that behaviour will be found in [ŽUK 1]. Even if this convergence was destroyed at each fifth step of the iteration by changing the formula (6.3.4) into

$$\hat{x} = x + (\alpha\omega)r, \qquad \omega > 1,$$

the remaining four iteration steps were sufficient to increase (6.3.5) over 0.9. This overrelaxation at each fifth step of the process results in a surprising acceleration of the eigenvalue convergence. Hence, we will describe a technique for a special construction of an overrelaxation diagonal matrix for the case $p > 1$ in order to change (6.3.3) in the same way to

$$\hat{X} = XQ_1 + RQ_2\Omega . \tag{6.3.6}$$

If Ω is introduced at each 5th step, then $\hat{X}$ will be no Ritz-tuple and the matrices $\hat{A}_{11}$ and $\hat{B}_{11}$ of the next iteration step must be calculated, that is, a restart is implied. Now, the following considerations aim at the construction of a matrix Ω , which must not disturb the convergence of the whole process. Our choice of Ω brought a considerable acceleration of the convergence.
In (6.3.6) $\hat{X}$ is no Ritz-tuple, hence we are able to simplify this formula to

$$\hat{X}_1 = X + RU. \tag{6.3.7}$$

For

$$U = U_1 = Q_2Q_1^{-1}$$

we would maximize $\mu(\hat{X})$ (Q contains eigenvectors of $\{A,B\}$ and is divided into the two blocks $Q_{1/2}$). For changing this choice at each 5th step we will use another $U = U_2$ which could be defined by maximizing another functional, for instance the absolute size of the projection of $\hat{X}$ onto span Φ_1. For further considerations we will deal with $\mathcal{B} = \mathcal{I}$ only. In this case this projection can be calculated and (approximately) maximized. Hence, we would use the functional

$$\chi(X) = tr(X^HX)^{-1}\ (X^H\mathcal{P}_1 X)$$

($\mathcal{P}_1$ the orthogonal projection operator onto span Φ_1) instead of

$$\mu(X) = tr(X^HX)^{-1}(X^H\mathcal{A}X).$$

The maximization of $\mathcal{X}(X)$ according to (6.3.7) over all $U \in \mathbb{R}^{p \times p}$ can be done with the help of differential calculus (compare 2.4) and would yield the result below. A more straightforward calculation of the element $\hat{X}_2$ uses the fact that, obviously, $\hat{X}_2$ with maximal $\mathcal{X}(\hat{X}_2)$ and (6.3.7) represents the best approximation of the columns Φ_1 in span $(X \vdots R)$. Thus

$$\hat{X}_2 = \mathcal{P} \Phi_1$$

where $\mathcal{P}$ denotes the orthogonal projection operator onto span $(X \vdots R)$,

$$\mathcal{P} = Y(Y^H Y)^{-1} Y^H$$

with

$$Y = (X \vdots R).$$

This leads to

$$\hat{X}_2 = (X \vdots R) \begin{pmatrix} X^H X & \mathbb{O} \\ \mathbb{O} & R^H R \end{pmatrix}^{-1} (X \vdots R)^H \Phi_1$$

$$= XH^T + R(R^H R)^{-1} R^H \Phi_1$$

with

$$H = \Phi_1^H X \in \mathbb{R}^{p \times p}.$$

Obviously,

$$R^H \Phi_1 = (\mathcal{A} X - X\Delta)^H \Phi_1 = (\Lambda_1 H - H\Delta)^T.$$

Hence, by changing the basis in span $\hat{X}_2$

$$U = U_2 = (R^H R)^{-1}(H^T \Lambda_1 H^{-T} - \Delta) \qquad (6.3.8)$$

would maximize $\mathcal{X}(\hat{X})$.

From Lemma 2.4.3 we conclude for U_1:

$$X^H R(\hat{X}_1) = \mathbb{O}$$

$$= X^H(\mathcal{A} \hat{X}_1 - \hat{X}_1 F(\hat{X}_1)) =$$

$$= \Delta + X^H \mathcal{A} RU_1 - F(\hat{X}_1) = \mathbb{O}.$$

Due to $X^H R = \mathbb{O}$ (therefore $X^H \mathcal{A} R = R^H R$) we have

$$U_1 = (R^H R)^{-1}(F(\hat{X}_1) - \Delta) \qquad (6.3.9)$$

Note that, here

$$F(\hat{X}_1) = (\hat{X}_1^H \hat{X}_1)^{-1}(\hat{X}_1^H \mathcal{A} \hat{X}_1)$$

and by (6.3.7)

$$\hat{X}_1 = \hat{X} Q_1^{-1} \quad \text{with } \hat{X}^H \hat{X} = I, \ \hat{X}^H \mathcal{A} \hat{X} = \hat{\Delta},$$

hence

$$F(\hat{X}_1) = Q_1 \hat{\Delta} Q_1^{-1}.$$

The final result is obtained from the comparison of U_1 in (6.3.9) and U_2 in (6.3.8):
The matrix

$$\tilde{\Omega} = (Q_1 \hat{\Delta} Q_1^{-1} - \Delta)^{-1}(H^T \Lambda_1 H^{-T} - \Delta)$$

would transform U_1 into U_2 or $\hat{X}_1$ into $\hat{X}_2$

$$\hat{X}_2 = X + RU_1 \tilde{\Omega} = X + RQ_2 Q_1^{-1} \tilde{\Omega}.$$

For a convenient application we will approximate the unknown matrix H. We recall that X represents a Ritz-tuple, which is near to Φ_1 for later steps of the iteration. Here, we have the fact that the non-diagonal entries of

$$H = \Phi_1^H X$$

are in the order of ε^2, if $G = \Phi_2^H X$ is in the order of ε (see [LONGSINE/McCORMICK 1, PARLETT 1]). Obviously, we approximate H by diag H. Due to the same fact we approximate Q_1 by diag Q_1, which yields a new easier formula for $\hat{X}$:

$$\hat{X} = XQ_1 + RQ_2\tilde{\Omega}$$

with

$$\tilde{\Omega} = (\hat{\Delta} - \Delta)^{-1}(\Lambda_1 - \Delta). \qquad (6.3.10)$$

This formula can be applied for defining a diagonal overrelaxation matrix Ω, if we, furthermore, approximate Λ_1 by the Aitken-technique analogous to 6.2. For instance,

$$\tilde{\omega}_j = \frac{\lambda_j - \delta_j}{\hat{\delta}_j - \delta_j} = \frac{\lambda_j - \delta_j}{(\lambda_j - \delta_j) - (\lambda_j - \hat{\delta}_j)} = \frac{1}{1 - \eta_j}.$$

From the fact that $\tilde{\Omega}$ in (6.3.10) has diagonal entries larger than 1, we recommend the final definition

$$\omega_j = \max\left(1, \frac{\delta_j - \check{\delta}_j}{2\delta_j - \hat{\delta}_j - \check{\delta}_j}\right), \quad j = 1,\ldots,p, \qquad (6.3.11)$$

$$\Omega = \operatorname{diag}(\omega_1, \ldots, \omega_p).$$

If at each 5th step the normal iteration formula

$$\hat{X} = XQ_1 + RQ_2$$

is changed to

$$\hat{X} = XQ_1 + RQ_2\Omega$$

with the diagonal matrix Ω according to (6.3.11), the resulting algorithm is fairly accelerated, especially at the consecutive step (each first normal step) the eigenvalue approximations jump. If the rise of one δ_j with (straight line) and without (dotted line) using these exceptional steps in the iteration is plotted, the following Figure 1 will show the typical behaviour.
We established the formulas for Ω by using $\mathcal{B} = \mathcal{I}$, but for the case $\mathcal{B} \neq \mathcal{I}$ the same behaviour occured as shown in the figure. Hence, we will also recommend this overrelaxation at the restart points for the algorithm PSGM(K,M,C).

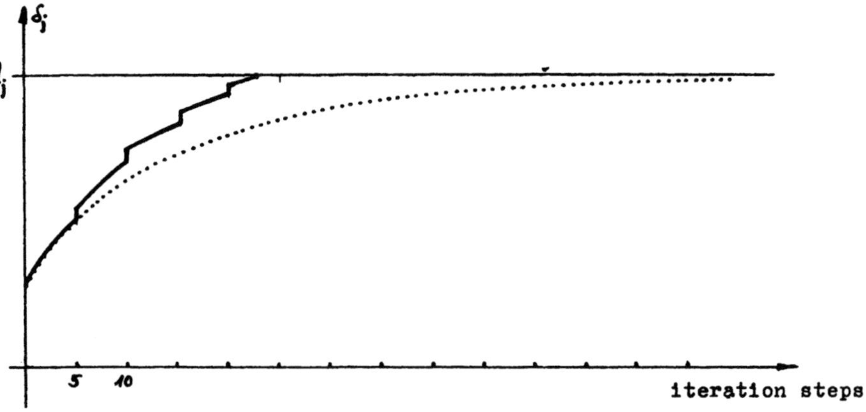

Figure 1

For the accelerated gradient methods of Chapters 3.2. and 3.3. such exceptional steps are not necessary. Here, the fact span $\hat{R} \approx$ span $\check{R}$ cannot be found. On the other hand, the use of the overrelaxation in the simple gradient method will often accelerate it near to the conjugate gradient or W-accelerated version, but is substantially cheaper than these methods.

REFERENCES

Alsen, B.M.

[1] Multiple step gradient iterative methods for computing eigenvalues for large symmetric matrices, Report UMINF 15.71, Univ. of Umeå 1971.

Axelsson, O.

[1] Solution of linear systems of equations: Iterative methods, In: Sparse Matrix Techniques, Copenhagen 1976, Lect. Notes in Math. 572.

Bathe, K.-J., Ramaswamy, S.

[1] An accelerated subspace iteration method, Comp. Meth. Appl. Mech. Engng. 23 (1980), 313-331.

Bathe, K.-J., Wilson, E.L.

[1] Solution methods for eigenvalue problems in structural mechanics, Int. J. Num. Math. Engng. 6 (1973), 213-226.

Bauer, F.L.

[1] Das Verfahren der Treppeniteration und verwandte Verfahren zur Lösung algebraischer Eigenwertprobleme, ZAMP 8 (1967), 214-235.

Clint, M., Jennings, A.

[1] The evaluation of eigenvalues and eigenvectors of real symmetric matrices by simultaneous iteration, Comp. J. 13 (1970), 76-80.

[2] A simultaneous iterative method for the unsymmetric eigenvalue problem, Int. J. Num. Math. Engng. 10 (1976), 647-663.

Corr, R.B., Jennings, A.

[1] A simultaneous iteration algorithm for symmetric eigenvalue problems, Int. J. Num. Math. Engng. 10 (1976), 647-663.

Cullum, J.

[1] The simultaneous computation of a few of the algebraically largest and smallest eigenvalues of a large, sparse, symmetric matrix, BIT 18 (1978), 265-275.

Cullum, J., Donath, W.E., Wolfe, P.

[1] The minimization of certain nondifferentiable sums of eigenvalues of symmetric matrices, Math. Progr. Studies No. 3, North Holland Publ., Amsterdam, 35-55.

Cullum, J. Willoughby, R.A.

[1] Lanczos algorithms for large symmetric eigenvalue computations, Vol. 1 and 2, Progress in Sci. Comp. Ser., Birkhäuser, Boston 1984.

[2] A survey of Lanczos procedures for very large real 'symmetric' eigenvalue problems, J. Comp. Appl. Math. 12 & 13 (1985), 37-60.

Davidson, E.R.

[1] The iterative calculation of a few of the lowest eigenvalues and corresponding eigenvectors of large real symmetric matrices, J. Comp. Phys. 17 (1975), 87-94.

Dietrich, G.

[1] On the efficient and accurate solution of the skewsymmetric eigenvalue problem, ISD Report No. 264, Stuttgart 1981.

Döhler, B.

[1] Ein neues simultanes Gradientenverfahren für das allgemeine großdimensionierte Eigenwertproblem, Dissertation, TH Karl-Marx-Stadt 1981.

[2] Ein neues Gradientenverfahren zur simultanen Berechnung der kleinsten oder größten Eigenwerte des allgemeinen Eigenwertproblems, Num. Math. 40 (1982)1, 79-91.

[3] Simultane Gradientenverfahren zur Maximierung des Funktionals μ(X), Wiss. Schriftenreihe TH Karl-Marx-Stadt 8 (1983), 57-79.

Döhler, B., Meyer, A.

[1] Optimale Gradientenverfahren für das spezielle Eigenwertproblem, Wiss. Schriftenreihe TH Karl-Marx-Stadt 8 (1983), 80-88.

Duff, I.S., Reid, J.K., Munksgaard, N., Nielsen, H.B.

[1] Direct solution of sets of linear equations whose matrix is sparse, symmetric and indefinite, J. Inst. Maths. Applics. 23 (1979), 235-250.

Eispack

[1] Matrix eigensystem routines - EISPACK guide, B.T. Smith, J.M. Boyle, B.S. Garbow, Y. Ikebe, V.C. Klema, and C.B. Moler, Eds., Lect. Notes in Comp. Science 6, Springer Verlag 1976.

[2] Matrix eigensystem routines - EISPACK guide extension, B.S. Garbow, J.M. Boyle, J.J. Dongarra and C.B. Moler, Eds., Lect. Notes in Comp. Science 51, Springer Verlag 1977.

Ericsson T., Ruhe, A.

[1] The spectral transformation Lanczos method for the numerical solution of large sparse generalized eigenvalue problems, Math. Comp. 35 (1980), 1251-1268.

Evans, D.J.

[1] Preconditioned iterative methods for the generalized eigenvalue problem, in: Matrix Pencils, Lect. Notes in Math. 973, 121-144.

Evans, D.J., Shanehchi, J.

[1] Preconditioned iterative methods for the large symmetric eigenvalue problem, Comp. Math. Appl. Mech. Engng. 31 (1982), 251-264.

Faddeyev D.K., Faddeyeva, W.N.

[1] Computational Methods of Linear Algebra, San Francisco, W. H. Freeman 1963 (also German: DVW, Berlin 1970).

Fan, Ky

[1] On a theorem of Weyl concerning eigenvalues of linear transformations, Proc. Natl. Acad. Sciences U.S.A. 35 (1949), 652-655.

Frach, K.-H., Greiner-Mai, H., Suschowk, G.

[1] Physikalische Experimente an Leistungsreaktoren, Akad. Verl., Berlin 1982.

Fried, I.

[1] Optimal gradient minimization scheme for finite element eigenproblems, J. Sound Vibration 20 (1972), 333-342.

[2] Bounds on the extremal eigenvalues of the stiffness and mass matrices and their spectral condition number, J. Sound Vibration 22 (1972), 407-418.

Geradin, M.

[1] The computational efficiency of a new minimization algorithm for eigenvalue analysis, J. Sound Vibration 19 (1971), 319-331.

Golebiewski, A.

[1] Variational pseudo-gradient method for determination first m eigenvalues of a large real symmetric matrix, in: Numerical integration of differential equations and large linear systems, Lect. Notes in Math. 968, Springer Verlag 1982, 370-383.

Golub, G.H., Underwood, R.

[1] The block Lanczos method for computing eigenvalues, in: Mathematical Software III, (Edited by J. Rice), Acad. Press, New York 1977, 361-377.

Gose, G.

[1] Relaxationsverfahren zur Minimierung von Funktionalen und Anwendung auf das Eigenwertproblem für symmetrische Matrizenpaare, Dissertation, TU Braunschweig 1974.

Gustafsson, I.

[1] A class of first order factorization methods, Computer Sciences 77.04R, Chalmers Univ., Goteborg.

[2] On first order factorization methods for the solution of problems with mixed boundary conditions and problems with discontinuous material coefficients, Computer Sciences 77.13R, Chalmers Univ., Goteborg.

[3] Stability and rate of convergence of modified incomplete factorization methods, Computer Sciences 79.02R, Chalmers Univ., Goteborg.

Hanzal, A.

[1] Varianten der Koordinatenüberrelaxation, Dissertation, Univ. Zürich 1978.

Jennings, A.

[1] The development and application of simultaneous iteration for eigenvalue problems, in: Conf. on Appl. of Num. Anal., Lect. Notes in Math. 228, Springer Verlag 1971, 297-309.

[2] Eigenvalue methods and the analysis of structural vibration, in: Duff, I.S., Sparse matrices and their uses, Acad. Press 1981, 109-138.

Jennings, A., Stewart, W.J.

[1] Simultaneous iteration for partial eigensolution of real matrices, J. Inst. Math. Appl. 15 (1975), 351-361.

Kahan, W., Parlett, B.N.

[1] How far should you go with the Lanczos process? in: Bunch, J., Rose, D., Sparse Matrix Computations, Proc. Symp. Lemont 1975, 131-144.

Korneev, V.G., Langer, U.

[1] Approximate solution of plastic flow theory problems, Teubner Texte 69, Leipzig 1984.

Kučerov, A.B., Makarov, M.M.

[1] The method of approximate factorization for the solution of difference equations of mixed boundary value problems (Russian), in: Difference methods in mathematical physics, Moscow Univ. 1984, 54-65.

Lanczos, C.

[1] An iteration method for the solution of the eigenvalue problem of linear differential and integral operators, J. Res. Nat. Bureau of Standards 45 (1950), 255-282.

Lezius, U.

[1] Spektraltransformationsalgorithmen für großdimensionierte Eigenwertprobleme, Diplomarbeit, TH Karl-Marx-Stadt 1985.

Longsine, D.E., Mc Cormick, S.F.

[1] Simultaneous Rayleigh-quotient minimization method for $Ax = \lambda Bx$, Lin. Alg. Appl. 34 (1980), 195-234.

Matthies, H.G.

[1] A subspace Lanczos method for the generalized symmetric eigenproblem, Computers & Structures 21 (1985), 319-325.

Mc Cormick, S.F., Noe, T.

[1] Simultaneous iteration for the matrix eigenvalue problem, Lin. Alg. Appl. 16 (1977), 43-56.

Meyer, A.

[1] Grundlegende Beziehungen für simultane Algorithmen, Wiss. Schriftenreihe TH Karl-Marx-Stadt 8 (1983), 3-30.

[2] Die Simultane Iteration und die simultane Tschebyscheff-Beschleunigung, Wiss. Schriftenreihe TH Karl-Marx-Stadt 8 (1983), 31-49.

[3] Die simultane Berechnung einiger Eigenwerte und Eigenvektoren beim großdimensionierten allgemeinen Matrixeigenwertproblem, Dissertation B, TH Karl-Marx-Stadt 1985.

[4] Numerische Software für Aufgaben der linearen Algebra, Teil II, IKM-Berichte 2, Weimar 1984, 27-29.

[5] Varianten der simultanen Iteration zur Berechnung einiger Eigenwerte und Eigenvektoren des allgemeinen großdimensionierten Eigenwertproblems, Beitr. Num. Math. 10 (1981), 107-115.

[6] Einige theoretische Aspekte bei der Simultanen Iteration zur Berechnung betragskleinster oder -größter Eigenwerte eines allgemeinen Eigenwertproblems, Wiss. Inf. TH Karl-Marx-Stadt 24 (1981).

[7] Die Simultane Iteration zur Eigenwertberechnung für beliebige endlichdimensionale Operatoren und die Anwendung auf Schwingungsprobleme, Wiss. Inf. TH Karl-Marx-Stadt 36 (1982).

Meyer, A., Döhler, B.

[1] Über simultane Eigenwertalgorithmen beim Schwingungsproblem, ZAMM 65 (1985), 597-604.

Neubert, M.

[1] Ein neues simultanes Lanczos-Verfahren für das Eigenwertproblem, Wiss. Inf. TH Karl-Marx-Stadt 56 (1985).

O'Leary, D.

[1] The block conjugate gradient algorithm and related methods, Lin. Alg. Appl. 29 (1980), 293-322.

Paige, C.C.

[1] Practical use of the symmetric Lanczos process with reorthogonalization, BIT 10 (1970), 183-195.

[2] Computational variants of the Lanczos method for the eigenproblem, J. Inst. Math. Appl. 10 (1972), 373-381.

[3] Error analysis of the Lanczos algorithm for tridiagonalizing a symmetric matrix, J. Inst. Math. Appl. 18 (1976), 341-349.

Parlett, B.N.

[1] The symmetric eigenvalue problem, Prentice-Hall, Englewood Cliffs N.J. (1980).

[2] The software scene in the extraction of eigenvalues from sparse matrices, SIAM J. Sci. Stat. Comput. 5 (1984), 590-604.

Parlett, B.N., Nour-Omid, B., Taylor, R.L.

[1] Lanczos versus subspace iteration for solution of eigenvalue problems, Int. J. Num. Meth. Engng. 17 (1983).

Parlett, B.N., Reid, J.

[1] Tracking the progress of the Lanczos algorithm for large, symmetric eigenproblems, IMA J. Num. Anal. 1 (1981), 135-155.

Parlett, B.N., Scott, D.S.

[1] The Lanczos algorithm with selective orthogonalization, Math. Comp. 33 (1979), 217-238.

Parlett, B.N., Taylor, D.R., Lin, Z.L.

[1] A look-ahead Lanczos algorithm for unsymmetric matrices, Math. Comp. 44 (1985), 105-124.

Prikazčikov, V.G.

[1] Prototypes of iteration processes for eigenvalue problems (Russian), Diff. Uravn. XVI (1980), 217-238.

Ruhe, A.

[1] SOR-methods for the eigenvalue problem with large sparse matrices, Math. Comp. 28 (1974), 695-710.

[2] Iterative eigenvalue algorithms based on convergent splittings, J. Comp. Phys. 19 (1975), 110-120.

Ruhe, A., Wiberg, T.

[1] The method of conjugate gradients used in inverse iteration, BIT 12 (1972), 543-554.

Rutishauser, H.

[1] Computational aspects to F.L. Bauers simultaneous iteration method, Num. Math. 13 (1969), 4-13.

[2] Simultaneous iteration method for symmetric matrices, Num. Math. 16 (1970), 205-223.

Saad, Y.

[1] Projection methods for solving large sparse eigenvalue problems, in: Matrix Pencils, Lect. Notes in Math. 973, Springer Verlag 1983, 121-144.

[2] Krylov subspace methods for solving large sparse unsymmetric linear systems, Math. Comp. 37 (1981), 105-126.

[3] On the rate of convergence of the Lanczos and the block-Lanczos methods, SIAM J. Numer. Anal. 17 (1980), 687-706.

Samarskij, A.A.

[1] Theorie der Differenzenverfahren, Akad. Verlagsgesellschaft Geest & Portig, Leipzig 1984.

Sameh, A.H., Wisniewski, J.A.

[1] A trace minimization algorithm for the generalized eigenvalue problem, SIAM J. Num. Anal. 19 (1982), 1243-1259.

Schwarz, H.R.

[1] Two algorithms for treating $Ax = \lambda Bx$, Comp. Meth. Appl. Mech. Engng. 12 (1977), 181-199.

[2] Simultane Iterationsverfahren für große allgemeine Eigenwertprobleme, Ing.-Archiv 50 (1981), 329-338.

[3] Simultaneous Rayleigh-quotient iteration methods for large sparse generalized eigenvalue problems, in: Num. Integr. of Diff. Equ. and Large Sparse Syst., Lect. Notes in Math. 968, 384-398.

[4] Die Methode der konjugierten Gradienten mit Vorkonditionierungen, in: Num. Behandlung von Differentialgleichungen Bd. 3, ISNM 56, Birkhäuser 1981.

Scott, D.S.

[1] The advantages of inverted operators in Rayleigh-Ritz approximations, SIAM J. Sci. Stat. Comp. 3 (1982), 68-75.

[2] How to make the Lanczos algorithm converge slowly, Math. Comp. 33 (1979), 239-247.

Shavitt, I.

[1] The method of configuration iteration, in: Methods of electronic structure theory, H.F. Schaeffer III ed., Plenum, New York 1977.

Simon , H.

[1] The Lanczos algorithm with partial reorthogonalization, Math. Comp. 42 (1984), 115-142.

Skurt, L.

[1] Das Verfahren der simultanen Koordinatenrelaxation, Wiss. Schriftenreihe TH Karl-Marx-Stadt 8 (1983), 50-56.

[2] Zur simultanen Koordinatengruppenrelaxation für das allgemeine Eigenwertproblem, Wiss. Inf. TH Karl-Marx-Stadt 41 (1983).

Stewart, G.W.

[1] A bibliographical tour of the large sparse generalized eigenvalue problem, in: Bunch, J., Rose, D., Sparse Matrix Computations II, Eigenvalue problems, proc. Symp. Lemont 1975, 113-131.

Stewart, W.J., Jennings, A.

[1] A simultaneous iteration algorithm for real matrices, ACM Transactions Math. Softw. 7 (1981), 184-198.

Underwood, R.

[1] An iterative block Lanczos method for the solution of large sparse symmetric eigenproblems, Ph.D. dissertation, STAN-CS-75-496, Stanford Univ.

Vandergraft, J.S.

[1] Generalized Rayleigh-methods with application to finding eigenvalues of large matrices, Lin. Alg.Appl. 4 (1971), 353-368.

Waldvogel, P.

[1] Numerische Behandlung von allgemeinen Eigenwertproblemen, Dissertation, Univ. Zürich 1985.

Wilkinson, J.H.

[1] The algebraic eigenvalue problem, Claredon Press, Oxford 1965.

Wilson, E.L.

[1] Numerical methods for dynamic analysis, in: Num. Meth. in Offshore Engineering, J. Wiley & Sons 1978, Ch. 6, 195-217.

Žuk, P.F.

[1] On asymptotical properties of the method of steepest descent for eigenvalue problems (Russian), J. Vyc. Mat. Mat. Fiz. 21 (1981), 271-285.

ANALYTIC PERTURBATION THEORY

FOR MATRICES AND OPERATORS

by

Hellmut Baumgärtel

Academy of Science of the GDR

Mathematische Lehrbücher und Monographien

II. Abteilung, Monographien, Band 64

1984. 427 pages, 17 by 24 cm, DDR 80,-- M, abroad 80,-- DM.

Order Number 763 352 6

Order word: Baumgaertel 6823 engl.

The aim of this book is to give a systematic presentation of the Analytic Perturbation Theory for Matrices and Operators, in particular on finite-dimensional spaces. This presentation closes essentially the topic by solving the basic open problems. Moreover, the consequences for analytic perturbations of isolated eigenvalues of finite algebraic multiplicity of closed operators on Banach spaces are discussed. The central questions are the development of the analytic perturbation theory for the spectral representation and for the Jordan structure of linear operators together with a reduction theory of analytic perturbations.

Methodically, the presentation is based on the spectral theory and the theory of Jordan normal forms (Jordan structure) for single operators and on a detailed discussion of the structure of linear operators on finite-dimensional linear spaces over arbitrary commutative fields.

In this book not only the perturbation theory for a single complex perturbation variable is presented but also the case of analytic perturbations depending on several complex variables.

The book contains numerous results of the author with full proofs previously unpublished. It is related to KATO's comprehensive monograph "Perturbation theory for linear operators".

The book is intended to be of interest to all those who have to make precise their ideas and their knowledge on analytic perturbation theory which plays a still growing role in many parts of Mathematics, Mathematical and Theoretical Physics.

AKADEMIE-VERLAG BERLIN

Band 1

Inverse and Improperly Posed Problems in Differential Equations

Proceedings of the Conference on Mathematical and Numerical Methods, held in Halle/Saale (GDR) from May 29 to June 2, 1979

Edited by Gottfried Anger
296 Seiten, 1979 (vergriffen)

Band 2

Fundamentals of Computation Theory FCT '79

Proceedings of the Conference on Algebraic, Arithmetic and Categorial Methods in Computation Theory, held in Berlin/Wendisch-Rietz (GDR) September 17–21, 1979

Edited by Lothar Budach
575 Seiten, 1979

Band 3

Decidability and Generalized Quantifiers

By Andreas Baudisch, Detlef Seese, Hans-Peter Tuschik, and Martin Weese
XII, 235 Seiten, 1980

Band 4

Zur Elastizitäts- und Thermoelastizitätstheorie

Potentialtheoretische Methoden in Anwendung auf verschiedene Modelle

Von Lothar Jentsch und Johannes Maul
157 Seiten, 1980

Band 5

Systems Analysis and Simulation 1980

Proceedings of the International Symposium held at Berlin (GDR) September 1–5, 1980

Edited by Achim Sydow
496 Seiten, 1980 (vergriffen)

Band 6

Nichtlineare diskrete Optimierung

von Eberhard Girlich und Michail M. Kowaljow
218 Seiten, 1981

Band 7

Completeness Problems in the Structural Theory of Automata

By Jürgen Dassow
148 Seiten, 1981

Band 8

Darstellung von Lösungen linearer elliptischer Differentialgleichungen

Von Günther Wildenhain
92 Seiten, 1981

Band 9

Introduction to the Spectral Theory of Operators in Spaces with an Indefinite Metric

By I. S. Iohvidov, M. G. Krein, and H. Langer
120 Seiten, 1982 (vergriffen)

Band 10

Nonlinear Time-discrete Systems

A General Approach by Nonlinear Superposition

By Michael Gössel
115 Seiten, 1982 (vergriffen)

Band 11

Preprimal Algebras

By Klaus Denecke
162 Seiten, 1982

Band 12

Ergodic Theory and Related Topics

Proceedings of the Conference held in Vitte/Hiddensee (GDR) October 19–23, 1981

Edited by Horst Michel
232 Seiten, 1982

Band 13

Singulariy Perturbed Differential Equations

By H. Goering, A. Felgenhauer, G. Lube, H.-G. Roos, and L. Tobiska
176 Seiten, 1983

Band 14

Leben wir in einer Volterra-Welt?

Ein ökologischer Zugang zur angewandten Systemanalyse

Von Manfred Peschel und Werner Mende
233 Seiten, 1983

Band 15

Markov Decision Problems with Countable State Spaces

Optimality Criteria – Algorithms – Clustering

By Hans Michael Dietz and Volker Nollau
174 Seiten, 1983

Band 16

Stochastische Verfahren zur Suche nach dem Minimum einer Funktion

Von Ryszard Zieliński und Peter Neumann
133 Seiten, 1983

Band 17

Invertibility and Asymptotics of Toeplitz Matrices

By Albrecht Böttcher and Bernd Silbermann
200 Seiten, 1983

Band 18

Structural Induction on Partial Algebras

Introduction to Theory and Application of Partial Algebras – Part II

By Horst Reichel
205 pages, 1984

Band 19

Algebraic Methods for Toeplitz-like Matrices and Operators

By Georg Heinig and Karla Rost
212 pages, 1984

Band 20

Frege Conference 1984

Proceedings of the International Conference held at Schwerin (GDR), September 10–14, 1984
Edited by Gerd Wechsung
408 pages, 1984

Band 21

Numerical Analysis of Selected Semilinear Differential Equations

Edited by Christian Großmann, Thomas Riedrich, and Hartmut Schönheinz
203 pages, 1984

Band 22

Singular Ordinary Differential Operators and Pseudodifferential Equations

By Johannes Elschner
200 pages, 1985

Band 23

Lotka-Volterra-Approach to Cooperation and Competition in Dynamic Systems

Proceedings of the 5th Meeting of UNESCO's Working Group on Systems Theory held on the Wartburg, Eisenach (GDR) at March 5–9, 1984
Edited by Werner Ebeling and Manfred Peschel
271 pages, 1985

Band 24

Convergence Structures 1984

Proceedings of the Conference on Convergence held in Bechyně (Czechoslovakia); September 24–28, 1984
Edited by Josef Novák, Werner Gähler, Horst Herrlich, and Jan Mikusiński
254 pages, 1985

Band 25

Parcella '84

Proceedings of the II. International Workshop on Parallel Processing by Cellular Automata and Arrays held in Berlin (GDR), September 25–27, 1984
Edited by Wolfgang Händler, Tamás Legendi, and Gottfried Wolf
239 pages, 1985

Band 26

Multiobjective and Stochastic Optimization Based on Parametric Optimization

By Jürgen Guddat, Francisco Guerra Vasquez, Klaus Tammer, and Klaus Wendler
175 pages, 1985

Band 27/28

Systems Analysis and Simulation 1985
I: Theory and Foundations
II: Applications

Proceedings of the International Symposium held in Berlin (GDR), August 26–31, 1985
Edited by Achim Sydow, Manfred Thoma, and Robert Vichnevetsky
per volume 495 pages, 1985

Band 29

Parcella '86

Proceedings of the III. International Workshop on Parallel Processing by Cellular Automata and Arrays held in Berlin (GDR), September 9–11, 1986
Edited by Tamás Legendi, Dennis Parkinson, Roland Vollmar, and Gottfried Wolf
about 272 pages, 1986

Band 30

Fuzzy Sets Applications, Methodological Approaches, and Results

Proceedings of the International Workshop on Fuzzy Sets Applications cosponsored by IIASA and convened in cooperation with IAEA held on the Wartburg, Eisenach (GDR), at March 3–8, 1985
Edited by Steffen Bocklisch, Sergei Orlovski, Manfred Peschel, and Yasushi Nishiwaki
233 pages, 1986

Band 31

Mathematical Methods of Specification and Synthesis of Software Systems '85

Proceedings of the International Spring School, held in Wendisch-Rietz (GDR), April 22–26, 1985
Edited by Wolfgang Bibel and Klaus P. Jantke
245 pages, 1986

Band 32

A Model Theoretic Oriented Approach to Partial Algebras

Introduction to Theory and Application of Partial Algebras – Part I
By Peter Burmeister
319 pages, 1986

Band 33

Finite-difference Methods on Irregular Networks

A Generalized Approach to Elliptic Problems of Second Order
By Bernd Heinrich
about 200 pages, 1986

Bei Fragen zur Produktsicherheit wenden Sie sich bitte an:
If you have any questions regarding product safety,
please contact:

Walter de Gruyter GmbH
Genthiner Straße 13
10785 Berlin
productsafety@degruyterbrill.com